Creating the New World

Stories & Images from the Dawn of the Atomic Age

Theodore Rockwell

Foreword by Glenn Seaborg, **Discoverer of Plutonium**

1st Books Library
Bloomington, Indiana

Second Edition

ISBN: 1-4033-9086-X (e-book)
ISBN: 1-4033-9087-8 (Paperback)
ISBN: 1-4107-0333-9 (Dust Jacket)

Library of Congress Control Number: 2002095507

This book is printed on acid free paper.

Printed in the United States of America
Bloomington, IN

1stBooks – rev. 12/02/03

"This is a wonderful account of the complex experiences of a thoroughly dedicated nuclear engineer. He relates his story with a freshness that brings back many memories to those of us who were also engaged in the enterprise in our own way. His saga makes it seem as though it had all happened yesterday."

DR. FREDERICK SEITZ
Past President, National Academy of Sciences

"Ted Rockwell has summed up more than a half-century of personal experiences as a pioneer in the nuclear age. His observations are poignant lenses on the key people and events that kindled and then nurtured the nuclear genies of electrical power, nuclear medicine, weapons, and radioisotopes in research and industry. His book is an important gift to this and coming generations."

Dr. JOHN H. GIBBONS
Assistant to the President for Science & Technology (1993-98)
Director, Congressional Office of Technology Assessment (1980-93)

"A very important, well-written, and easily read book. Ted Rockwell's part in the development of atomic energy began during the Manhattan project and continues to the present day. With this background he sharply challenges many assumptions that restrict the development of nuclear power and the beneficial uses of radiation."

Dr. FRANK DUNCAN
U.S. Atomic Energy Commission historian

"In the last few years, unnoticed yet significant developments in nuclear technology have begun to dispel public disapprobation of all things nuclear. Ted Rockwell is a knowledgeable, articulate communicator with the courage born of a lifetime of successful nuclear achievement, practical as well as theoretical. His book *Creating the New World* may well be a critical element in further enlightening our fellow citizens to the important and many benefits of nuclear science and technology."

Dr. KENNETH C. ROGERS
U.S. Nuclear Regulatory Commissioner (1987–1997)
President Emeritus, Stevens Institute of Technology
Fellow, AAAS and of American Nuclear Society

OTHER BOOKS BY THEODORE ROCKWELL

The Reactor Shielding Design Manual (editor). Published separately in 1956 by U.S. Government Printing Office, McGraw-Hill, and Van Nostrand. Widely republished, translated and excerpted abroad. Still used as text.

The Shippingport Pressurized Water Reactor (co-author). Part of U.S. contribution to 1958 Geneva Conference on Atomic Energy. Published by Addison-Wesley. Cited by American Library Association as "one of the best technical books of 1958."

Arms Control Agreements: Designs for Verification (co-author), Johns Hopkins Press (1968), used in connection with US-USSR talks at the White House.

The Rickover Effect: How One Man Made a Difference, Naval Institute Press (1992); Chinese language edition (1994); John Wiley paperback (1995) republished by the Authors Guild through iUniverse, Inc., 2003. Excerpted in Reader's Digest, domestic and foreign editions.

Vice Versa: Three One-Act Plays, professionally performed as a staged reading by Source Theater, Washington, DC (1990).

Some Articles by Theodore Rockwell

"Frontier Life Among the Atom Splitters," *Saturday Evening Post*, Dec 1, 1945.

"Bred For Fury," *True*, July 1946. (First color stroboflash photos of fighting cocks in action).

"Water Chemistry of Pressurized Water Reactors" (co-author). Presented at First International Conference on Atomic Energy (1955); published by United Nations and by Pergamon Press.

"Heresy, Excommunication, and Other Weeds in the Garden of Science." *New Realities*, Dec 1981.

"Report of Panel on Science and Unexplained Phenomena," *American Society for Information Science*, U.S. Bicentennial Conference, Apr. 12-14, 1976.

"Irrational Rationalists," (Senior author). *Jour. Amer. Soc. Psychical Research*, 72, 23-34, (Jan 1978). Reprinted in *The Battlefield of Psi* (Japanese language anthology, 1987).

"Admiral H.G. Rickover's Impact on Nuclear Power." Invited opening paper, Chairman's Special Session, presented Nov 16, 1992 at the Joint International Conference of the American Nuclear Society and the European Nuclear Society, commemorating the 50th Anniversary of the Birth of Nuclear Power.

CONTENTS

FOREWORD

This book is primarily a collection of first-hand stories, images and essays about the development of nuclear power for the production of electricity. It presents in vivid, human terms many of the young scientists and engineers who first harnessed this primal force and the extraordinary times and environment in which they worked and lived. As a bystander and at times a participant in this development, I have found this to be an enlightening and fascinating account. Included is an absorbing insider's description of the origin and build-up of our nuclear navy, a process that contributed so much to civilian nuclear power. I have had the unusual advantage of knowing both the author, Ted Rockwell, and his redoubtable mentor, Hyman Rickover, over much of this time.

Nuclear power today produces 20% of the electricity in the United States and a similar amount throughout the rest of the world. The end of the cold war has made large quantities of nuclear fuel from dismantled weapons for use in reactors potentially available, and the development of breeder reactors should provide essentially unlimited nuclear fuel for the future.

Our fossil fuels—wood, coal, oil and gas—are running out, and they are needed as raw materials to make medicines, plastics, synthetic fibers, dyes and a myriad of the basics of modern life. Burning these to produce electricity creates huge quantities of waste products, such as acid rain, air pollution and greenhouse gases that heat up our fragile planet. Advocates say we have unlocked the limitless power of the atom just in time, and countries like France and Japan are taking advantage of this.

However, the growth of nuclear power in the United States and some other countries has come to a stop, and opponents of nuclear power in the United States and elsewhere project an adverse picture for its use in the future. They depict as insolvable the problems of radiation exposure, nuclear safety, and radioactive waste disposal. In *Creating the New World* the author deals with these problems in a reasonable, rational manner to show that a better understanding by the general public is the key to the eventual acceptance of this available and safest source of increasingly needed electric energy.

The last chapter is written in a philosophical vein in an attempt to convince the lay reader that sufficient understanding of scientific and technical jargon is possible to an extent needed to form reasonable and positive judgments on these questions about nuclear energy.

Creating the New World should be successful in providing a human interest background and making a convincing case for nuclear power for the interested reader.

GLENN T. SEABORG
Nobel Laureate, Co-discoverer of plutonium
Chairman, U.S. Atomic Energy Commission, 1961-71
Chair-Emeritus, National Science Talent Search

"The Italian Navigator has landed in The New World," said Compton.
"How were the natives?" asked Conant.
"Very friendly."
Nobel laureate A.H. Compton telephoning Harvard President James Conant
that Enrico Fermi had demonstrated the world's first controlled nuclear chain reaction.

Preface

On December 2, 1942, the brilliant Italian physicist Enrico Fermi and his colleagues discovered and awakened the nuclear dragon—the primordial energy that binds the atom together and lights the sun and all the stars. Dr. Crawford Greenewalt, Chairman of DuPont and a witness to that historic event, agreed to accept President Roosevelt's challenge to help fashion a harness for the fearsome beast, putting some of DuPont's best onto the project, for cost and a one-dollar profit. And by December 1943, Westinghouse, General Electric, Stone & Webster, Tennessee Eastman, Union Carbide and other engineering giants were already erecting mammoth structures across the verdant Smoky Mountains. That's when I first went down to the new and secret city, Oak Ridge, Tennessee, to join in the engineering effort to convert the physicist's dream to a decisive weapon and an energy source to fuel The New World.

Theodore Rockwell
May 2002

Acknowledgements

For my wife, Mary Compton Rockwell, who endured much during the seven years gestation it took to bring this baby to term. She provided not only sustained encouragement but also her artist's insights as to what was needed. And to my children, Bob, Teed, Larry and Nita, artists all, in their own way. Each made special contributions, both as to general tone and focus, but also many specific ideas as to content. And to the people at the Writer's Center and at Washington Independent Writers, whose contributions during several workshops helped me to hone my writing skills. Graphics are a key ingredient in shaping this book's character. Each of the drawings credited to "Hoke" was specially drawn for this book by J. Nevin Hoke and his group of talented designer/draftsmen at MPR Associates. The contemporary photographs of Oak Ridge credited to "DOE" were taken by J. Edward Westcott, who was the official photo chronicler of Oak Ridge from the time the bulldozers first bit into the red and yellow clay. He not only took marvelous photographs, but fifty years later was able to reach into the files and pick out just what I needed. I can't imagine trying to write this book without both of these unique artists. And special thanks to Roxanne Fournier Stone, editor extraordinaire, whose skill and dedication wrought a raw manuscript into a book. Thanks to you all!

——Ted Rockwell

Creating the New World

1. Nuclear Genesis

Facing the Beast

"Ever see an atomic bomb go off, Rockwell?"

"No, sir."

"You want to?"

"Well, yeah. Sure."

My enigmatic boss, then-captain Hyman Rickover, the legendary "father of the nuclear Navy," looked back at his papers, signaling that the conversation was over. I had no idea why he asked. I filed the conversation fragment away in my mind along with a lot of other such cryptic exchanges and went back to work. But several months later, after I had nearly forgotten about it, I received official Navy orders dispatching me on 20 March 1953, to a hot tent at Camp Desert Rock, Nevada, to spend four days getting ready to experience first-hand the detonation of an atomic "device," as these demonstration weapons were called.

I was one of a half-dozen civilians who would be cowering in a hand-dug slit-trench at Yucca Flats, with a regiment of three thousand troops in full battle gear, only four thousand yards from the largest bomb ever detonated in the continental U.S. (*Four thousand yards! And I thought observers were always several miles away—and in concrete bunkers!)* Many of the soldiers were survivors of Normandy, Guadalcanal, or Anzio. Their fatigues, net-covered steel helmets, packs and rifles were well used; they had been trained in the gas masks they wore. There was no protective clothing or equipment for us civilians, so we wore whatever we had come with, in my case a red-and-black plaid shirt and gray work pants. I didn't even have a hat or a jacket. We had official orders "to participate as observers," but we did not figure in any of their plans. The battle plan simulated a hostile attack on Las Vegas. We had been allotted nine atomic weapons in the scenario, although the actual test would involve only one real device.

The shot was scheduled for 5:10 A.M. March 23, 1953, and reveille came at 11:16 P.M. the night before, or 2316 hours in Army lingo. The timing wasn't casual; it had been written down and passed around days ago. We spent the next six hours eating breakfast and generally milling around and waiting. With my simple civilian mind, I couldn't understand why a seemingly arbitrarily scheduled event had to start at sixteen minutes after anything, let alone after eleven at night. I'm the kind of person who just doesn't wait well. All my life I've crammed activity into every waking minute; to suddenly have nothing to do but *wait* was a shock to my system. The military, however, were used to it.

After standing for an hour and a half in the cold, dark, slit-trench, the silence was suddenly broken by an announcement: a one-ton World War II blockbuster demonstration shot would be fired at about the same distance from the trenches as the atomic test shot, to give a comparison standard. I was startled by the sharpness of the THUMP on my chest, and I listened solemnly as the loudspeaker blared impersonally: "The actual shot will be an estimated thirty-five thousand, repeat, thirty-five thousand times as powerful as the one you have just witnessed." *Wow!* I thought. *TWICE the wallop of the bomb that incinerated Hiroshima!* "You are warned that to expose yourself during this shot may be fatal. Keep your head down until the audible blast has passed."

I was shivering in the cold pre-dawn air, waiting for something to happen, when the countdown began. On my knees, burying my face in my thighs, I was intensely aware of the smell of the dry dirt and rock, and the silence of the desert night. In spite of myself I began to feel real fear as the countdown continued with the inevitability of the voice of doom. *That S.O.B. can afford to sound calm, I thought. He's at the observation post, ten miles away. Did the clowns who calculated I'd be safe really know what they're doing? I don't see any of them here. They can't even be sure how big a bang this will be. This is really just an experiment. I'm somebody's guinea pig. Did I really volunteer for this? I didn't even hear the count ZERO!*

But all thought ceased when the unearthly white light began to intensify. Slowly, in deadly silence it grew. Not yellow-white, not blue-white, but white, white, white. The whitest, brightest light I had ever seen. I was staring at the rocks between my knees, and the shadowless light on them seemed to come from everywhere, growing in the eerie silence as if nothing could stop it until the entire world was consumed. "Brighter than a thousand suns," the physicists said, and certainly it seemed so. But what made it seem so unreal was the complete silence, as the shock waves from the explosion raced toward me through the dry Nevada tuff. I felt no boom, no shock, no thump to the chest, just that bright, silent light.

I was a numb observer of all this—present, but still an observer. Suddenly I became a terrified participant as the solid earth beneath me and the walls of my protective slit-trench turned to jelly. Only four seconds of stillness had elapsed, during which I realized I had been holding my breath; but for the next eight seconds, I'll never know what I did. At that moment I fully understood the terror of earthquake victims as the very earth, the foundation of all being, dissolved beneath my feet. And still the deathly silence persisted.

Finally, because sound and shock travel five times more slowly through air than through the earth, the airborne shock wave and the sound of the blast smote our trenches. The suddenness of it was physically breathtaking. Dirt and sand and pieces of twigs and other debris rained down on me, and for a moment I was afraid of being buried alive. *What the hell am I doing here?* I cried. I didn't know—or care—whether I only thought it in the depths of my mind or shrieked it aloud. The initial thunderclap of the long-awaited sound did not die out, but incredibly it continued to increase and reverberate off the mountains with the same majestic deliberation that had characterized the light. I don't know whether the light continued to grow with the sound, or whether it blacked out altogether. Any message from my eyes to my brain was lost in the frantic S.O.S. from all my other senses.

After what seemed like hours but was probably seconds, I realized I had been burying my head in my arms for some time. I started to raise my head, ever so cautiously. I saw some of the soldiers standing full upright, and I sheepishly stood to look around. The sun was blotted out by smoke and debris, and the Joshua trees were burning. The poles holding the loudspeakers were charred. The roots of the sage, exposed as the cyclonic winds blew the sandy soil away, were smoking or burning softly in the eerie light. A singed bird skittered crazily through the gloom. Even without the mushroom cloud, it looked like a painting of hell. My first impression of the cloud was: *How close it is!* It towered forty thousand feet into the sky and looked as if it were almost directly overhead. Myriad hues of pink, lavender, red, and yellow were flickering in the cloud. I was surprised that it was still generating light, and way up there! I recalled that J. Robert Oppenheimer, seeing the first A-bomb detonated at Alamogordo, had reportedly quoted from the Bhagavad-Gita, "Now I am become Shiva, destroyer of worlds."

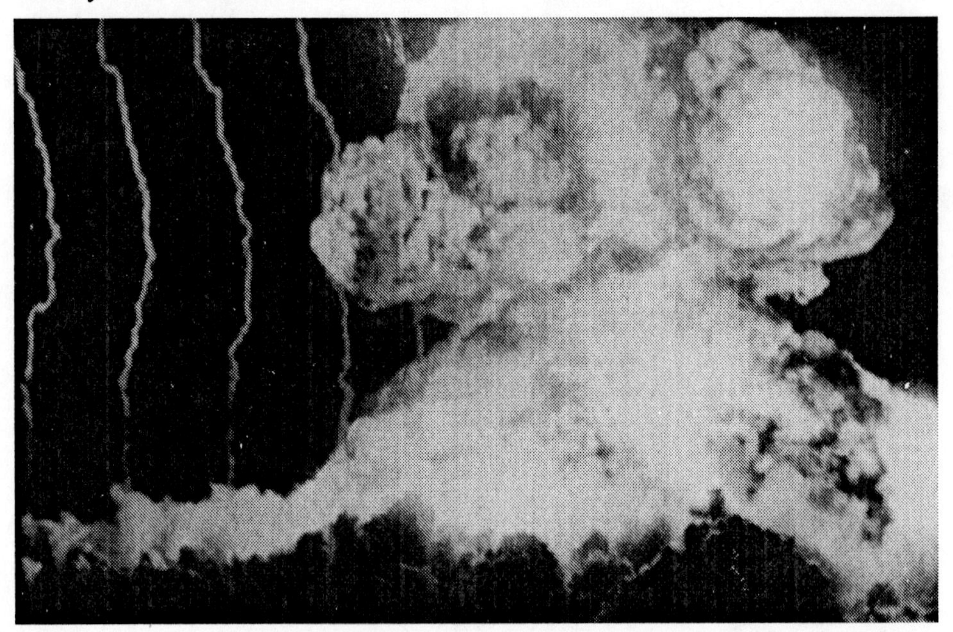

Figure 1.1 The atomic fireball from 4,000 yards (National Archives)

I finally emerged from my trance to see the troops out of the trenches and advancing in full battle gear toward ground zero. I scrambled out to join them. Although the sun was coming up, the turgid smoke and dust from the smoldering desert darkened the hellish landscape more and more as we moved in. The thick cloud billowing up below the mushroom had started to settle uncertainly toward the trenches. The loudspeaker announced, "This will be the first time that friendly troops have been caught in the base surge." *Friendly troops, hell*, I thought. *That's me he's talking about.*

We walked past the civil defense demonstrations in which test dummies were seated in various types of houses. These had been set up for the previous bomb test to show the public how to protect themselves. But they became so radioactive that no one had been allowed near them. We kept walking—fast. When we finally got back to the loading area to return to camp, we were brushed off with ordinary kitchen brooms as radiation monitors scanned for residual contamination on our clothing. The experience was over, but I would not soon forget it.

Figure 1.2 Troops climb from their slit-trench and prepare to move in toward Ground Zero (*National Archives*)

The soldiers, on the other hand, were not about to admit being shaken. *Newsweek* quoted a number of them, each trying to be cooler than the one before:

> "I didn't think it was worth coming out here for."
>
> "It would make a fine tactical weapon, and that's all."
>
> "Charge six of one-five-five would hurt your ears more."
>
> "It wasn't bad."

I thought back seven and a half years, to the day when the news of the fury of the atomic bomb was first released to the world, and Hiroshima suddenly became an American household word.

"OAK RIDGE ATTACKS JAPANESE"

It was mid-morning, August 6, 1945, in Oak Ridge, Tennessee. I had been there a year and a half, and the wartime work pace had been kept at the limit the whole time. The day shift had been at work since 7:30A.M.; a minimum 6-day, 54-hour base work week had become the norm. Although Oak Ridge wasn't on any maps, it was the fifth largest city in the state, with a population of 75,000, and an additional 40,000 workers commuting from outside the Site. It was a real live city with schools, theaters, library, hospital, thirteen supermarkets, buses, a telephone system, a newspaper, recreational facilities, restaurants, churches, houses and dormitories. It fenced in 59,000 acres of Roane and Anderson counties, but for quite a while, the Governor of Tennessee didn't even know it was there. Harry Truman heard of it the first time only four months previously, when he became President. It was built secretly from scratch in about two and a half years by the Manhattan District Project of the Army Engineer Corps, for the sole purpose of producing material for a fearsome new weapon. Few of its citizens knew this at the time; all they were told was they were producing some kind of secret substance, sometimes called a "catalyst," that would end the war. And that was enough to keep them working long and hard.

At the production building where I worked, at a location known only as "Y-12," one of the men got a call from home, which usually meant some kind of emergency; wives didn't call at work to chat. He hurried to the phone with a worried look. "*What?!* Say that again! Look, you'd better not say anything. Wait 'til we get more information."

Oak Ridge Attacks Japanese

STAY ON THE JOB

OAK RIDGE JOURNAL

U. S. POSTAGE SEC. 562, P. L. & R. PAID Oak Ridge, Tenn. Permit No. 1

Vol. 3—No. 4 OAK RIDGE, TENNESSEE Thursday, August 9, 1945

Workers Thrill As Atomic Bomb Secret Breaks; Press And Radio Stories Describe 'Fantastically Powerful' Weapon; Expected To Save Many Lives

Teamwork Responsible, Nichols Says

Teamwork was the outstanding factor in the accomplishments of CEW, Col. Kenneth D. Nichols stated at a press conference this week. The scientists started their their work first, but the laborers who worked on the roads and the thousands of workers who cooperated so amazingly deserve much credit.

Organization was the keynote in accomplishing the job quickly and safely, and the results of the CEW Safety program alone are outstanding. The 1944 employe injury rates of the plants were less

To Contractors, Workers, And Residents Of Oak Ridge:

CONGRATULATIONS to all workers at the Clinton Engineer Works and to the people of Oak Ridge! You have done the impossible.

I am sure that you shared with me the thrill which came with President Truman's announcement that the results of our hard work and American "stick-to-it-iveness" have been delivered to the Japs with a world-shaking crash—the thrill that comes with the knowledge of a tough job well done.

This project has been, from the start, a cooperative enterprise, based on mutual faith—faith of the scientist that engineers could translate his discoveries—yes, and his world stirring dreams—into practical process designs; faith of the engineer that material and construction men could turn those designs into bricks and mortar and process equipment; faith of the Army that all contractors would have the vision, courage, and drive to do the seemingly impossible; faith of the operating contractors that local nontechnical workers could be trained to perform new and strange tasks so exacting that they would normally be entrusted only to skilled scientific experimenters; faith of

Bomb Has More Power Than 20,000 Tons Of TNT; Pride Is Reaction Of Workers

The most exciting day in the era of secrecy and suspense which has governed the life of Oak Ridge residents occurred Monday with President Truman's announcement that an "atomic bomb" had been used against the Japanese, and the subsequent stories in the press and on the radio describing the part of the CEW in the production of the fantastically powerful weapon.

The reaction of Oak Ridge residents and workers on the whole was a feeling of pride that they had a part to play in this vital development, and a thrilling satisfaction in knowing that their efforts had been so effective, as the story of the efficacy of the bomb was told.

Stating that the bomb had more power than 20,000 tons of TNT, the President described it as "harnessing of the basic

Look Inside . . .

. . . for stories on the following subjects:

COL. NICHOLS:
. . . In the News, Page 2.

OFFICIAL STORIES:
. . . of Oak Ridge and background, with pictures, Pages 3-4-5.

NEW DORM MANAGERS:
. . . New Firm takes over dormitories, Page 6.

HOSPITAL:
. . . New business policy at hospital, Page 7.

SOFTBALL:

Figure 1.3 The August 9, 1945 *Oak Ridge Journal* "OAK RIDGE ATTACKS JAPANESE!" (*Author*)

He came back and looked at us, started to say something, but saw that one of the men had a lower security code on his badge, and stopped short. Other phones started ringing, and then the shift supervisor spoke out on the office intercom. "President Truman has just announced on the radio that an atomic bomb has been dropped on a Japanese city. It's been wiped off the map. He didn't say much else. Something about harnessing the basic power of the universe. And they did mention Oak Ridge. But I don't think anybody ought to say anything until we see what's been released."

There should have been a great release of emotion at this point: our mission had been successfully completed; the end of the war was in sight. There was, however, the simultaneous realization that we had unleashed an awesome genie. Later on, there was a reaction—but not yet. It was too soon. There was a lot of whispering, and people looking over their shoulders as they tried to carry out their normal duties. A short time later, people started coming back from lunch, bringing newspapers with them. And there it all was: OAK RIDGE SECRET BARED; U.S. USES ATOMIC BOMB and OAK RIDGE ATTACKS JAPANESE. WORKERS THRILL AS ATOMIC BOMB SECRET BREAKS. My favorite was simply: OAK RIDGE HAS OVER 425 BUILDINGS. One of the newsboys kept repeating, "Read all about Oak Ridge. We used to hate 'em, now we love 'em." One article noted solicitously: "'There's *no danger* of an atomic blast at Oak Ridge,' commanding officer reveals."

There was a terrible ambivalence about the atom. The stark reality of Hiroshima and Nagasaki, and the realization that Washington, London and Paris could be next, was a paralyzing experience. But it was equally real that huge American invasion forces had assembled for the anticipated final bloody assaults on the Japanese homeland. *Operation Olympic,* scheduled for Kyushu in November 1945, and *Operation Coronet* for Honshu the following March, were suddenly not needed and were soon gratefully demobilized. We were told that the planned invasions of the Japanese home islands would have resulted in an estimated one million American casualties plus unprecedented destruction of life and property among the Japanese.

In addition, we never knew until VE Day whether the Germans, who had discovered nuclear fission, would beat us to the goal. American intelligence had turned up a memo written December 16, 1944, from Dr. Walther Gerlach, head of the German atomic bomb program, to Martin Bormann on Hitler's staff, boasting, "I am convinced that we are at the present time considerably farther ahead of America, both in research and development." Samuel Goudsmit, the eminent editor of *Physical Review*, was a Dutch émigré who knew most of the important German physicists and had been sent abroad by President Roosevelt to find out what he could about the status of the German program. He told me he was staying with British friends when the V-1 missiles began to fall. The local people responded with classic British calm, "... but I was in panic," he said. "I *knew* that they were just getting the range, and the next one would contain the atomic warhead! And I couldn't even tell them why I was so scared!"

The Threat and the Promise

In the first aftershocks following Hiroshima and Nagasaki, the atom seemed to threaten unprecedented wholesale destruction on the one hand, and to promise limitless energy for a peaceful and prosperous world on the other. We in Oak Ridge felt a special responsibility to see that the people and the policymakers understood the full reality of what had happened and what it meant. We felt *The Bomb* was so horrible as to make another war unthinkable. We intended to make that fact a positive motivator in national and international thinking, and within a few days scientists and engineers throughout the atomic community started moving to convert our strong personal feelings into group action. In the post-war euphoria, it seemed perfectly natural to us that we could—and should—do this.

It appeared that everything was now out in the open. But letters sent the day after Hiroshima to each employee and signed by the Under Secretary of War warned, "No one of you has worked on the entire project or known the whole story ... keep the secrets you have kept so well. The need for security and for continued effort is fully as great now as it ever was." We didn't know that General Groves was, at that very moment, informing a shocked President Truman that the U.S. had no additional bombs whatsoever. The two bombs used on Japan had constituted our entire stockpile. By that time, however, production of weapons-grade material had finally gotten into gear, and fuel for additional bombs was becoming available in rapidly increasing quantities.

With so much detailed information in the newspapers, it was hard to know what to keep secret. A significant step to answering that question appeared in the newspapers the following Sunday. The War Department announced that a complete report to the public, in the form of a hardback book published by the Princeton University Press, would be made available immediately. It was written by Professor Henry DeWolf Smyth, who was my freshman physics teacher and later U.S. ambassador to the

International Atomic Energy Agency in Vienna. The book explained the atomic fission process and the steps taken at Oak Ridge by the Manhattan Project to separate the fissionable isotope of uranium from the much more abundant non-fissionable part. It described the process of producing the new fissionable element plutonium at another huge and secret installation called Hanford, in the state of Washington. And it told about the Los Alamos Laboratory in New Mexico, which was designing and assembling the bombs. The book was replete with diagrams, pictures and tables, and sold for two dollars. Ten thousand copies were allotted to the Oak Ridge area.

Publication of the Smyth Report was to most of us at Oak Ridge, a bold and unexpected stroke. We could certainly agree with Smyth's closing words: "The people of the country must be informed if they are to discharge their responsibilities wisely." But the psychological shock from seeing years of total secrecy shattered in a matter of hours was overwhelming. I avidly read everything in the papers, but couldn't keep from starting to cover up the paper, hiding the page when someone approached. It was like being caught reading a SECRET document in the town square! I just couldn't reconcile seeing words like *uranium* and *atomic* and even *plutonium* right out there in the open, with lots of uncleared people around.

People went into Townsite or the Rec Hall or other public places, just to mill around, listen to the chatter, and occasionally make some inane remark to people they didn't know. At some point, I noticed that church bells were ringing. The newspapers and radio commentators kept referring to "harnessing the power of the sun and the stars," and I overheard a local citizen declare knowingly to a companion, "I *knowed* it was something about the stars, 'cause they had all them arc welders." That afternoon I rode home from work with a normally low-key division director, who kept shouting out the window to no one in particular: "We're making uranium!" Everyone was feeling proud and important, except for one sad-faced little lad who had to keep telling people, "My daddy is only a dentist."

Figure 1.4 The author and other Oak Ridgers celebrate the end of the war (*DOE*)

Oak Ridge had been remarkably calm on VE Day when the war in Europe was won. I don't recall any celebrations. We had been told repeatedly that this development would not be significant for us: the Japanese were still at war and we could not slack off. Toward the end of June 1945, there was a "Win the War in July" campaign, and we were told to scrape together every bit of product we could muster during that month. A number of us had a very private betting pool on when the war would end. Nearly all of us guessed it would cease between October and the end of the year. On August 6, we were astounded that uranium salt, which left Oak Ridge during the last few days of July, had been made into metal, fabricated into a weapon, shipped to an air base on Tinian, and dropped on Japan in such a short period. I am still dumbfounded at the thought. After that, VJ Day and the actual end of the war were something of an anticlimax.

How It Came About

By now, the media were bombarding us with stories about Dr. Einstein's simple equation showing how much energy is released when a small amount of matter is destroyed. *Matter destroyed?!* cried a generation of science teachers. *We were all taught that matter cannot be created or destroyed! How did this all come about?*

Figure 1.5 Wonderful possibilities of the Atomic Age envisioned (*Author*)

In 1939, Albert Einstein wrote to President Roosevelt that German scientists had discovered that when uranium atoms were bombarded with neutrons, they could be split in two—fissioned—releasing a great deal of energy in the process. In addition, it appeared that more neutrons were released with each fission, suggesting the possibility of a self-sustaining chain reaction causing other atoms to fission. These findings had been published in open scientific journals and had been promptly confirmed in universities and in public and private research institutions around the world.

It appeared that a weapon of terrible magnitude might be created from this process, and the Germans were presumably working to develop such a weapon. In view of that, Einstein recommended that American scientists explore this situation. Roosevelt quickly set up a Uranium Committee outside the existing bureaucracies, which recommended that the subject be developed on an urgent and highly classified basis. At first, a number of universities carried out the work—Columbia, Chicago, Princeton, UCal Berkeley, et al—but it soon became clear that a major program was required. The program was, therefore, assigned to the Army Corps of Engineers, code-named the Manhattan Project. The Army drafted PhD scientists as privates, and had its contractors scouring the country for civilian scientists, engineers, technicians and laborers, who began building whole communities in the wilderness, all in strictest secrecy.

One such community was Oak Ridge.

The Manhattan Project was a vast, $2 billion project, in the days when you could mail a letter for three cents. (Fifty years later, the city of Washington, DC, proposed spending more than $2 billion to repair the Woodrow Wilson Bridge.) The Manhattan Project involved thousands of organizations and hundreds of thousands of individuals. A hundred-volume series of books by the Atomic Energy Commission summarizes some of its technical accomplishments. I was able to see only a glimpse of this huge enterprise, but what I saw was fascinating and unforgettable. There is no substitute for being there while the action is on.

First Glimpses of the Secret City

In the fall of 1943, I had just turned 21 and was interviewing for a job while finishing up my chemical engineering graduate work at Princeton. I found the interviewer from the Tennessee Eastman Corporation particularly intriguing. He said he was offering an important war job, but he wouldn't tell me where it was or what it would be. He just said he was sure I would find it challenging and important. He finally admitted it was somewhere near Knoxville in east Tennessee. Nosing around, I located a chemist friend who was doing a secret research project. He wouldn't tell me anything but suggested guardedly that I look at recent issues of the chemistry journals. The only item of interest I could find was an article reporting detection of trace amounts of barium and strontium in uranium bombarded by neutrons. I was not sure what that signified, but when an almanac stated (erroneously, I found years later) that uranium ore had been discovered in east Tennessee, I jumped to the tentative conclusion that they might be working on atomic power. I looked up the recruiter the next day and asked him outright, "It's atomic power, isn't it?" He looked flustered and said, "We never confirm or deny speculations as to the nature of secret war projects." I took this as a confirmation of my wild guess, and immediately accepted his offer to go to the plant for an interview. The next day I was visited by a government security officer telling me with unambiguous clarity that I'd better keep my speculations to myself. I got the message.

Since my master's thesis at Princeton was an urgent classified project connected with the national wartime program to develop a synthetic substitute for natural rubber, I did not want to take any time off until Christmas break. So I scheduled my interview for that time and was told to report to the Andrew Johnson Hotel in Knoxville, which I did. After some milling around and some paper processing, I joined the other job applicants, climbing into a special type of Oak Ridge machinery, the "stretch car," which looked like something out of a Laurel and Hardy movie. It was a '39 Chevy, sawed in half, with a metal-and-plywood section inserted. I hoped that some extra bracing had been added! The lines of the vehicle were buried in reddish brown mud, dripping from the fenders, clotted under the running boards, and smeared across all the windows. Only the windshield, scraped by a wiper that ran rain or shine, provided a dim view of the outside world.

After a wild ride over bumpy country roads, we stopped at a military inspection post, where the car and luggage were searched and our passes and papers examined. Entering the portal, we strained to catch a glimpse of the secret city, but it remained safely hidden behind the coated windows. After another abrupt stop, the driver announced, "Person-nayull. All out." The interviews themselves were routine; I liked the people, but I learned nothing more about the job. I returned to Princeton to await the verdict and finish my graduate project. In due course, I received a letter offering me a job with a salary but no job description, and I accepted.

Figure 1.6 Oak Ridge citizen ponders implications of atomic energy (*Dave Robbins*)

The following April (1944), I returned to the same temporary-looking structure in Oak Ridge to be "cleared" and "processed." After filling out vast numbers of forms, I was assigned a dormitory room designated merely "WV-33-154," which meant room number 154 in the 33rd dormitory in West Village, a residential area about two or three miles down the road. I went to the bus "ternimal," as it was often called, and found to my delight that all bus service in Oak Ridge was free. The trip to West Village introduced me to another unique vehicle, a huge trailer bus. The cab was a typical truck unit, but the trailer was unlike anything I had ever seen—an Army olivedrab, cattle-car sort of cabin with wooden benches and two tiny windows: one up front, looking into the back window of the cab, and one in the single door on the side. After a mile or two, I could see through the little window a cluster of shacks that looked almost like platform tents, creating the appearance of a Klondike gold miners' camp. Some of these shanties had signs designating them as "General Store," "Laundry," and the like, while others appeared to be living quarters. The bus stopped, and I concluded from the dress of the passengers who got off that this was one of the living areas for construction workers.

The bus took off again and there was one more abrupt stop, after which the bus turned, proceeded a short distance, backed up, and turned the other way. An experienced passenger chuckled and announced that excavation for a new building had apparently begun in the middle of the roadway, and the driver was trying to improvise a new path through the muddy fields. I learned later that new houses were being built at the rate of one every thirty minutes. No wonder bus drivers and postmen often stopped to ask directions from anyone walking by. Finally, the bus stopped in front of two rows of H-shaped two-story wooden buildings, and the passengers assured me this was West Village. A year later, I would come to admire the graceful curved walks, grass, badminton courts, outdoor dance pavilions and canteens, and rustic wooden bridges that created a pleasant ambiance. But now, all I could see was an apparently aimless array of wooden buildings, dwarfed by the gargantuan chunks of red and yellow clay pushed up by bulldozers and front-end loaders.

The mud and dust were an experience all by themselves. I had never before seen a place where you could sink slowly into albuminous red muck while choking in a thick cloud of heavy, yellow dust. I learned that it was standard Oak Ridge etiquette to remove one's shoes before entering a house, and the dormitories usually had a line of people in front of the janitor's sink waiting to wash their shoes under the faucet. Mud was an accepted part of life. The conservative, well-pressed suit, with a heavy layer of mud on the bottom, was right in fashion. The ladies, bless 'em, seemed to float above such earthly contamination. Their eternal spotlessness was one of the inspirations that pulled us sodden males through that first dismal spring. The women soon learned to arrive at the dances immaculately dressed, remove their mud-caked hipboots, and leave them in the foyer with the others. Then they slipped on a pair of golden sandals and tripped lightly into the rough-hewn "Rec Hall" to dance to the recorded music of Glenn Miller and Benny Goodman.

Figure 1.7 Mud presents a formidable obstacle to construction of first Oak Ridge roads and houses (*DOE*)

The Frontier

The wartime Manhattan Atomic Bomb Project bears about the same relationship to the subsequent nuclear power program as the opening of the American frontier does to the later development of towns and cities. Pioneers have different talents and interests from those who adapt the new terrain for human use and benefit. Pioneers live in a different historical setting and generally move on to let others settle and develop the new territory. Both the pioneers and those who follow demonstrate their own brand of creativity, tenacity and competence; neither is superior to the other. But to understand fully the later events, you have to get at least a glimpse of the pioneering activity from which they grew.

Many historians and sociologists have written about the unique impact on the American psyche of the continuing existence of The Frontier—new people moving into land long occupied by a different people and a different culture, driving those people out and replacing them with their own ideas and artifacts. The land is torn up ruthlessly to create crude new towns, and people visiting from settled, "civilized" communities are repelled by the upheaval, the primitive living conditions, and the sheer rawness of it.

The new frontiersmen invaded Black Oak Ridge, driving out families who had lived there for generations. Many were ordered to leave their family homes on a few days notice, abandoning crops and animals in the fields, tobacco curing in the barn, and graves of their ancestors. As they looked back, the displaced families saw new security fences being installed, blocking their return. Some of them had been driven off their farms only a decade before, as the Tennessee Valley Authority (TVA) with its dams and hydroelectric power plants moved in only a few miles away. There were even some who had been moved to carve out the Great Smoky Mountain National Park a few years before that. They were not of a mind to recall that their own ancestors had driven out the Cherokee people in an almost identical manner, when the American frontier was a new experience for the nation.

The displacement procedure was cold and merciless, and the reimbursements rendered were pitifully inadequate. But one aspect was handled with sensitivity, and that was the graveyards. Seventy small cemeteries were preserved intact within the reservation, fenced off with the same barbed wire the displaced families had used to border their farms. In the center of town, Georgia Avenue swerves suddenly to the right before it enters Oak Ridge Turnpike to avoid a small family plot with six tombstones, carefully preserved under the original six large cedars and some newer dogwoods and mimosas. Special arrangements were made to enable people to visit once a year and decorate their family graves, and this practice continued for as long as the area was fenced off from the public.

Figure 1.8 Former residents of the area are admitted once a year to tend family graves (*DOE*)

Many writers romanticize the "natural" life, close to the land. But frontier people do not look back nostalgically to more primitive days; they look forward to a better world they feel they are helping to create. They are willing to endure frontier conditions as long as necessary, so that their children and their children's children may live better. They recall the words that John Adams wrote in 1780 to his wife Abigail: "I must study politics and war that my sons may have liberty to study mathematics and philosophy [science] ... in order to give their children the right to study painting, poetry, music, architecture, statuary, tapestry and porcelain." (My own grandchildren seem interested enough in music, but I have yet to get them involved in statuary, tapestry or porcelain.)

Life Inside the Fence

I found you couldn't wander far in any direction in Oak Ridge without having a guard come up and ask you what you were doing there. "Just looking around, getting the lay of the land," didn't do it, so you confined your wanderings to a few well-trodden paths. Wives and children living on a ridge could look out their bedroom windows at night and see the glow in the sky above the next ridge. The roar of construction machinery signaled the presence of the work areas that they could never visit. The lights and the noise had driven away the birds and demolished any difference between night and day, between workweek and weekends. Work went on continually here. At my dormitory there was a never-ending poker game. People pulled out at all hours, usually when they were ahead, but always with the excuse, "Gotta go to work." And the game went on, day in, day out.

The first school year at Oak Ridge had just begun, with an enrollment of 637 students. By the end of that school year, enrollment had risen to 5,000. The first baby was born on the reservation just before I arrived. She was born nine days before the hospital was opened and was named Elizabeth Ann. Her last name was not reported, for security reasons. We may have had a backwoods town, but we used 20 percent more electricity than New York City, and we had the sixth largest bus system in the nation. Our teachers came from forty different states, and our superintendent of schools came from Columbia University. And despite the shortage of materials, we soon had 163 miles of boardwalks made from scrap lumber from construction sites, winding picturesquely through the forests.

Oak Ridge was unusual in other ways as well. There were no extremes of rich or poor. The fanciest houses in town were all pre-fab look-a-likes. No amount of rank or pull could get you a better one. But there were no slums and no homeless or unemployed people. It was a city of young people; there were always weddings and baby showers, and it was a long time before the town could support a mortuary. This also meant it was a city without traditions or established social structures. There were no grandparents, no aunts and uncles, no extended families, until these were built up out of new arrivals to this fresh, new community. But you could feel the raw energy all around you as the recent immigrants worked at constructing not only roads and buildings, but social and cultural infrastructure as well.

The natives, some still speaking almost Elizabethan English despite the influx of TVA “flatland furiners” a decade before, found themselves faced with a bewildering invasion of high-tech machinery and a variegated assortment of scientists and engineers. They came from many cultures and backgrounds, but almost all were highly urbanized. But all were dependent on the ability of young women operators—many still in their teens—recruited from local countryside to operate their specialized machinery. Somewhere, in the bits of free time that popped up, people started babysitting exchange circles, book review groups, film classics clubs, Recording for the Blind, and a symphony orchestra. However, none of these clubs was allowed to affiliate with any national organization until after the war. Security did not want any more indications than necessary of the existence of this city or the names and number of its inhabitants.

Government regulation created a certain kind of class segregation. Top-paid officials were assigned the better houses and tended to socialize together, not only because they were neighbors, but also because they had the same degree of clearance for classified information; they were in less danger of saying something they shouldn’t. Construction workers were assigned their own areas. Young scientists usually lived in the dorms. But in the unique manner of American education, the children of all classes came together in the same schools. Children of Nobel laureates, of plant managers, and of day laborers, sat in the same classrooms. They came from New York City, from the mid-west cornfields, from the suburbs of California and from the local hills. Their parents had been recruited from the Alcan highway project in Alaska, from a closing magnesium plant near Las Vegas, and from a trade school near Pittsburgh, with ads that read: “When You’re a Grandmother, You’ll Brag About Having Worked at Tennessee Eastman.” And the kids learned as much from their peers as they did from their books.

I was single during most of my time at Oak Ridge. The atmosphere for single professionals was almost like a university. Most of my close associates and I lived in dormitories with assigned roommates who were also recent college graduates. As in pre-war colleges, we even had paternalistic management acting *in loco parentis*. No women, not even mothers or wives, were allowed past the front desk in the men's dorms. Some married women, on the waiting list for a house, were actually arrested for sneaking into their husbands' dorm rooms. Dorm dwellers lucky enough to be invited to a friend's house for dinner reciprocated the favor by smuggling a roll of scarce toilet paper from the dormitory bathroom.

Many eminent professors were on loan here, and special courses, seminars and workshops were in session at all times, on topics ranging from quantum mechanics to the electron theory of metals. There was a "student shop," where experienced machinists taught young scientists how to turn out items not readily available commercially. Distribution lists on reports had more Ph.D.s than misters. I found the presence of so many bright, dedicated young scientists a constant inspiration.

I had been in Oak Ridge for more than two weeks, but I had not seen a work area. I had seen buses labeled "Y-12," "K-25," "S-50" and "X-10", but I was not allowed to board them. We tried to figure out what these code names meant; were they coordinates on a map? We finally decided that German and Japanese intelligence would waste a lot of time on this exercise, and that the names were probably completely arbitrary, just to keep them guessing. (This turned out to be the case.) I spent my time on "The Hill," as the Oak Ridge headquarters area was called, awaiting military clearance. I attended lectures on Tennessee driving laws, first aid, plant safety, and Tennessee folklore, and saw an occasional archaic Felix the Cat or Mickey Mouse cartoon. Some of the "students" were obviously brilliant and highly educated; some were actually illiterate and had to get a classmate to write their names for them. The roll was called before every class, and each day a few lucky persons whose clearances were complete were released for work. Then one day, I heard my own name called, and a new phase of my life began. I would report to Y-12, a 12,000-acre plant site operated by the Tennessee Eastman Corporation, several miles southeast of the townsite, over East Fork Ridge and Pine Ridge, in the "great buildings and factories" of Bear Creek Valley, foretold by John Hendrix forty years previously.

2. The Manhattan Project Mystique

The Prophesy

A hundred years ago, old John Hendrix lay grief-stricken on the hard ground of Pine Ridge, near the Great Smoky Mountains of Tennessee. His stepchild had died, and John felt old beyond his four-score years. What could the future hold for him now? As he lay there in despair, he heard a Voice from the sky, ordering him to leave his family and go into the wilderness and meditate and pray. After forty days and forty nights he saw a glowing mist, and in the mist a vision, and he stumbled out of the wilderness to tell his people what he had been shown:

> *I've seen it coming ... Bear Creek Valley some day will fill with great buildings and factories and they will help win the greatest war that ever will be. There will be a city on Black Oak Ridge and I say the center of authority will be middleway between Tadlock's farm and Pyatt's place. A railroad will spur off the main L&N, run down toward Robertsville and then branch off toward Scarbrough. It will serve the great city of my vision. This I know.*
>
> *Big engines will dig big ditches....Thousands of people will run to and fro. They will build things and there will be great noise and confusion and the earth will shake.*
>
> *I've seen it. It's coming...*
>
> (Prophesy quoted from *Abiding Appalachia: Where Mountain and Atom Meet,* by Marilou Bonham Thompson)

His words were recorded but little heeded, and the people of Bear Creek Valley and Black Oak Ridge went back to their chores and their isolated lifestyle for another forty years.

The town and residential areas sparkled with the many-colored ID badges representing the scores of companies that worked on the site, but at Y-12 there was only one. The town badge did not admit one to the secured work area. Rather, each day I exchanged my town badge at the plant gate for a blue cardboard tag with my name, employee number and some capital block letters on it.

Figure 2.1 Security badge contains considerable information

In July 1944, I was issued a plastic picture badge with the Roman numeral IV, showing what level of classified information I was entitled to receive. The Arabic numeral 3 showed what work phase I was involved with, and the capital letters C, D, H, and K signified what fenced-in work areas I could enter. Payroll numbers were assigned sequentially, so you could tell whether another employee was hired before or after you. My number, 8403, came up in December 1943. So, overall, the badge gave a pretty complete picture of a person's place in the scheme of things.

A Roman V on the badge meant they would tell you everything. You didn't see many of those. My "IV" meant I was told all the technical details of the Y-12 process, but nothing about other parts of the Manhattan Project. That was just fine with me.

With the badge, I was given a memo that said, in part:

> The responsibility of having unrestricted access to process and development information is a very serious one.... Information such as you will have is given to only a few, and entails added responsibility...

Inside the plant site, I was taken to the training school where a remarkable and interesting program unfolded. First, it was clear that they were not going to tell the operators what was being produced at the plant; the product was merely referred to as a catalyst. Second, all parts of the equipment were given code names and/or letters. At first, this seemed like an insurmountable obstacle to learning the job, but it quickly became clear that it was actually a help. A letter was assigned to each specific part of the equipment, such as a metal casting, an electronic circuit, an evacuated tank, a magnetic field or an electric plasma. Chemicals were assigned three-digit numbers. Buildings were tagged with four-digit numbers, followed by "dash one" or "dash two," denoting its place in a particular series.

In addition, as in any plant, the workers dreamed up slang terms for everything and anything, and these names were standardized and encouraged. They formed an admirable code system that seemed natural to learn and to use, but extremely difficult for an outsider to figure out. For example, a wedge-shaped piece of carbon was used to "scrape" the stream of uranium-235 ions from the heavier uranium-238 ions in the plasma, and these units were dubbed *shavers*. A group of shavers, mounted in an assembly, formed a *barbershop*. Each design version was required by custom to have a proper name to distinguish it from later models, and so the first barbershop design was naturally called *Figaro*, after the premier barber of Seville. Exotic terms such as *ceramic shadowing*, *electrostatic prism* and *magnetic shim* referred to cheap blocks of material, while a fabulously valuable chemical solution was callously called *exhausted impoverished gunk. Gunk,* of course, was a very specific chemical, not to be confused with, say, *Crud*, which was another. *Exhausted* and *impoverished* referred to the results of particular steps in the process, and even words like *hot* and *cold* referred in specific instances to chemical concentration, electric potential or degree of radioactivity.

Figure 2.2 Guards routinely search cars at gates to Oak Ridge residential area (*DOE*)

The environment created by high-level classification was also new to me. My thesis work at Princeton for the National Research Defense Council had been stamped RESTRICTED, but there was little paperwork associated with that level of classification. At Oak Ridge I found double envelopes, sealing wax, receipts in triplicate, special couriers, red wastebaskets for scratch paper (emptied by armed guards each night and burned in the presence of military security guards) and all the other accouterments of a special secret project in wartime. On April 23, 1944, a welder was shot making an illegal entry. It seems he had been called at home in the middle of the night to come in for an emergency job. He made the long, tedious drive to the site and was told there was no pass for him. A few phone calls revealed that the pass was waiting at another gate. The welder was not about to go all the way back to Knoxville just to take another road leading to the opposite end of the site. "This is an emergency job. I'm going in," he said. The guard countered, "If you do, you're a dead man." He did, and he was. Knowing how hard it was to get skilled welders, our reaction was, *if they'll shoot a welder, they'll shoot anybody.*

Security was so tight that persons' names were often kept secret. Only players' first names were used in the local paper's reports of early high school football games. The *Philadelphia Inquirer* reported after the war: "Death certificates of employees accidentally killed on the Project were classified and weren't delivered to next of kin until after the war." Some of the more famous scientists had *noms de guerre*: Dr. Niels Bohr was "Mr. Nicholas Baker," Arthur Holly Compton was "Dr. Holly," Ernest Lawrence was "Ernest Lawson," Fermi was "Farmer," Wigner was "Wagner," and so on.

Most people weren't aware of the surreptitious code name system, and this often created strange situations. My wife's maiden name is Compton, so when her sister was introduced to a Mrs. Compton one evening, she naturally started to inquire as to her family background. The woman flushed, turned and walked quickly away. After the war, she learned about the use of code names and realized that her hostess, in making the introduction, had momentarily forgotten to apply Dr. Compton's code name to his wife.

All cameras, binoculars, telescopes and firearms had to be registered with local authorities. Signs warned "ENEMY EARS ARE NEAR YOUR BEERS" (without regard for the difficulty of getting beer). My favorite was a poster featuring Groucho Marx in his then-popular talk show role, with the caption: "SAY THE SECRET WORD AND GET $10,000 AND/OR TEN YEARS IN JAIL." These warnings took on added meaning when we were told that one out of every four adults was a government informer. And we didn't know who they were. (This cloak of anonymity was rent somewhat when a team identifying themselves as FBI entered our local baseball league.) We learned that people who asked too many questions or talked inappropriately disappeared from our midst. We didn't ask what happened to them, but we assumed they were returned to the outside world, with strict orders not to mention anything they had seen.

This extraordinary security program proved to be remarkably effective. Apparently, neither the Japanese nor the Germans ever really learned much about the Project, and their scientists and military people were astounded by the news of Hiroshima. Japan had virtually no atomic weapons project, and Germany, which discovered atomic fission, did not make any significant headway in its efforts to create an atomic bomb.

A friend named Mac Barrett told me of an unsettling incident. He wrote a letter to his former landlady in Berkeley, California, and dropped the letter in a sidewalk mailbox in Knoxville, since he happened to be there shopping. Technical correspondence between Oak Ridge and Berkeley was handled through a special mail drop, so that the postmarks would not reveal the linkage between the two places. My friend did not use the mail drop for this particular letter, thinking that it might attract more attention than a normal letter. When he returned to his room, he found the letter slipped under his door, unmailed, with a note telling him to use the mail drop. Such incidents unnerved him, but he found the answer after he married the woman he went to Knoxville with, and she told him she had been working for the FBI.

Yet there were many advantages to our peculiar way of life. We never locked our houses or our cars. Milkmen would come into the house and put the milk directly into the "ice box," as I called it in those days. Children played safely in the woods and parks. With the rest of the Nation, we endured rationing of sugar, meat and gasoline, but our houses had hardwood floors, brick fireplaces, and new electrical appliances unobtainable elsewhere. Electricity and water were free, as was trash pick-up. Soft coal was delivered free to the coal bin in each house, but the frequent coal dust explosions in our furnaces posed a constant challenge to the cleanliness of the white organdy "glass curtains" that were so common. The most important advantage, of course, was that wives and children had their men living with them, and the men were appreciative of being able to do work they liked instead of being shot at overseas.

Town Planning

Although it was not obvious at first, the town was fortunate in having an exceptional architectural team from Skidmore, Owings and Merrill. They were not told where the town was to be located. Initially they were told to plan for 500 residents, but that figure was revised sharply upward every few months or so. Once construction got underway, they were given a requirement of 12,000 residents. By fall of 1943, this number had jumped to 42,000. By January of 1945 the figure had grown to 66,000, and eight months later the actual population was over 75,000.

Figure 2.3 First houses spring up on the hillsides (*DOE*)

There was a broad range of housing. At the top of the scale were several thousand houses for senior scientists and operating personnel. There were “A houses” and “B houses”—all the way up to “F houses.” Six different styles of family houses were built of two-by-fours and cement-and-asbestos sheeting, with red brick fireplaces. They designed the kitchen and bedroom facing the street, to shorten the distance for plumbing and utility lines. This enabled the big picture window and porch to look out into the shady backyard, often shared by other houses on the circle and featuring a barbecue or other amenities. For $38 to $73 a month, including coal delivery, water, electricity, and garbage collection, we felt we were getting a bargain. To make these rent numbers meaningful in today's terms, note that when I got married after the war and moved into a Type E apartment, the rent and included services was about 10 percent of my base pay. Each dormitory room for two single persons was 12 x 16 feet. It had two beds, two nightstands, two desks with lamps, and two chests of drawers, all made of simple unpainted wood, but new. For all this we paid $7.50 a month (each) for rent.

In addition to the dormitories for unmarried workers and barracks for the Army's enlisted men, there were also apartments for families. As the demand for housing outran the supply, several hundred prefabs called flattops, originally designed for TVA, were brought in on trucks, already completely furnished. When these ran out, more Spartan prefab housing, called victory cottages were hauled in.

Figure 2.4 Construction camp trailers (*DOE*)

Housing for construction workers was considerably more primitive and was located in separate areas, away from the "permanent" housing. After completion of the work, the construction communities were torn down and the areas converted to other uses. The large construction companies provided some trailers, and the Army brought in thousands more from other installations around the country. These were basically just bedrooms with central bathhouses and laundry facilities located down long, muddy paths.

Primitive as they were, trailers were better than the *hutments*, which were sixteen-foot-square buildings made of quarter-inch plywood nailed to two-by-two studs. With no insulation, they were stifling in the summer, cold in the winter. Hutments had no windows and one door. They had a bed in each corner and a pot-bellied stove in the center. At the peak of Oak Ridge's settlement, 32,000 workers and their families lived in barracks, trailers or hutments.

Figure 2.5 The hutment area (*DOE*)

African-Americans at Oak Ridge

Although FDR's Executive Order 8802 in 1941 had officially prohibited discrimination in defense industries, America's southland was still segregated. Initial plans for Oak Ridge, based on a total population of 2,500, envisioned a model "Negro Village," segregated from the white community but composed of the same type of houses, dormitories, cafeteria, church, and a few stores. But as the Army revised the Oak Ridge total population ever upward, the better-class housing was assigned on the basis of job title. Since African-Americans were essentially all in the lower-paying jobs, Negro Village became East Village, another community for whites. It was rationalized by some that black people were not accustomed to such fancy housing and would probably not have been comfortable there anyway.

Eventually there was a separate community for the 1,500 black workers, who comprised about 2 percent of the total workforce, but it was all in dirt-floored hutments. The women lived in a women's compound, separated from the men by a five-foot fence topped with barbed wire. An armed guard enforced the ten o'clock curfew. It was reported that sometimes the guards pulled men off the fence, bleeding from trying to reach their wives. No black children were allowed on the reservation; in fact, there was no family life whatsoever for African-Americans. If a black woman became pregnant, she lost her job and was put outside the gate of the city to wait for a bus.

Although their living conditions were scarcely better than off the reservation, they were tolerated because the pay of fifty-eight cents an hour and more was considerably better than many of them had been able to get "outside." That sounds ridiculously low today, but for reference, my starting salary there, with two engineering degrees, was $48 per 40-hour week or $1.20 per hour. It should be further noted that about 4,500 white construction workers were also living in hutments, though not in fenced-in compounds. Another 10,000 lived in barracks. Ironically, some of these barracks dwellers were among the highest-paid people on the site, getting time-and-a-half or double-time for long work hours.

After the war, things improved for the black people, but only slowly and as a result of pressure from some of the scientists who had finally learned about the appalling living conditions. As one who was completely oblivious to this situation until many years later, I now begin to understand my contemporaries in Germany, whose children demanded, “How could you not be aware of what was happening to the Jews? How could you not know?” Today it seems strange, but at the time I just didn’t think about the way the black workers were living. We were heavily compartmentalized. Not only did I not see their quarters, I never visited any of the construction workers’ living areas, or the Army barracks, or any other production or support facilities on the reservation. We were not permitted to wander around, seeing or asking about how others worked or lived. People who asked too many questions were fired as security risks.

Figure 2.6 Wartime workers leaving Y-12 plantsite at shift change (*DOE*)

Women, too, suffered under the patriarchal attitude of the times. There were a very few female scientists, some of whom made important technical contributions, as well as some carefree young single women. But most women came to Oak Ridge, not for their own purposes, but to live as their sisters did “outside,” cooking and housekeeping for their men and raising their children. Their isolation was made more painful by the security regulations that prevented their husbands from talking about the work, which occupied nearly all their waking hours.

The Great Silver Caper

Despite the tight restrictions that the war had imposed on all construction, vast quantities of materials poured into Oak Ridge. In one eleven-week period, 38 million board-feet of lumber, 5 million bricks, and 13,000 windows were delivered to Y-12. One particularly scarce material was copper, and Y-12 needed a lot of it to carry the heavy electric currents to the great magnets used in the separation process. But ingenuity prevailed. Colonel Ken Nichols, the brilliant 34-year old Army Ph.D. engineer in charge of the entire Manhattan Project under General Groves, learned that 47,000 tons of pure silver (yes, *tons*) was available to war projects. He quickly commandeered 14,700 tons of it from the U.S. Treasury's West Point Depository to be fabricated into strips 5/8-inch thick, 3 inches wide, and 40 feet long. These became the "wires" that fed electricity into the magnet coils. They in turn were fed from huge bars of solid silver about a square foot in cross-section running around the top of the entire circumference of the "race track," as the production units were called. After the War, the silver bars were disassembled and returned.

The job continually presented new experiences. When I walked the wooden catwalk above the production units, a strange force tugged at the nails in my shoes, giving a feeling of walking through light glue. After my initial confusion, I realized that the one thousand huge magnets that bent the glowing beams of ionized uranium atoms into curved paths, permitting the isotopes to be separated, had a magnetic field that extended to where I was walking. Non-magnetic tools had to be used in this area. Occasionally, a worker would carelessly get too close with an ordinary wrench or hammer, and it would fly out of his hand, crashing against the tank wall. Watchmakers in Oak Ridge got used to workers bringing in watches whose innards were smashed but whose cases had never been opened. (Watchmakers in other cities couldn't believe it.)

I also quickly learned a sobering lesson in Oak Ridge priorities. One day a crew was rigging a huge steel faceplate for a vacuum test tank. It was suspended from a large crane, and the crew inadvertently got it too close to the magnet. Sweeping into the magnetic field, it suddenly crashed against the magnet, pinning a workman against the magnet face. There were shouts to shut off the magnet, but the grim-faced building manager refused to do so. "Shutting off the magnet will take the whole track out of production until we can start up again and get everything back to equilibrium. Maybe a day or two, maybe more."

Figure 2.7 Billboard reminds Oak Ridgers of their mission (*DOE*)

The crew looked at him in disbelief. He didn't waver. "They tell us the war is killing 300 people every hour. And we're making the stuff that's supposed to stop the war. Whatever has happened to this guy has already happened. You'll just have to pry up that faceplate with two-by-fours. Gently! I'm not shutting off the magnet." The workers moved quickly to free their comrade who, remarkably, was not badly injured.

On another occasion, I was looking through the heavy lead glass inspection window at the beautiful blue glow of the uranium plasma, when the tungsten heat shields suddenly shone white-hot and then melted and ran in rivulets down the copper structure. Copper and silver-solder vaporized in a lavender flash, the whole system was aglow, and the unit shut down, all within a few seconds. An electron oscillation had just flared up. Copper fins had been designed into the equipment to prevent such oscillations, but it was found that the electrons quickly burned through them. The next step was to install refractory tungsten and tantalum inserts to resist the heavy electrical currents that sometimes built up during these transients. This usually worked, but not always, as I could now personally testify.

Perhaps the most bewildering phenomenon was the operators of these powerful devices. Most of them were young women, who had not graduated from high school, and were trained for the job under a wholly false description of the process and its purpose. Particularly to the young urbanized physicists from New York City, these "hillbilly girls," with their Daisy Mae accents and casual attitudes seemed totally unaware of the awesome responsibility entailed in operating these sophisticated machines. As I looked from the molten chaos to the placid face of the operator, I couldn't escape a feeling that the forces of Nature that had submitted to the minds brilliant enough to conceive these devices, might yet rebel at yielding to the will of someone totally uncomprehending of the energies involved. And yet, I had seen these same operators develop an intuitive sensitivity to the equipment, like a country boy and his ancient pickup truck, and had seen them get a unit purring again after a Ph.D. scientist had botched an effort to improve performance. These women personalized their units, and they complained bitterly if they were forced to operate another, although identical, unit. When some of the production buildings were shut down after the war, the operators wrote sentimental notes and farewell poems with lipstick on their instrument panels. They shed real tears of sorrow at the forced separation from the esoteric devices with which they had developed such a strange and productive relationship.

At the time, I was bemused by all this and didn't take it seriously. I eventually learned that such intuitive wisdom was not only useful to semi-literate mountain girls, but was also an essential ingredient in any creative scientific or technological endeavor.

The Y-12 Process

All atomic elements are made up of several *isotopes*. The isotopes for each element are all identical in terms of chemical properties, but each has a slightly different weight. The fissionable isotope of uranium is called

U-235 because its weight is about 235 times that of the hydrogen atom, which has a weight of approximately one on the atomic scale. Natural uranium, as it is dug out of the ground, has over a hundred atoms of U-238 for every atom of U-235. U-238 is not fissionable and is, therefore, useless for bombs. The Y-12 process separates out the valuable U-235 from the useless U-238. (U-238 has other important uses, but that's another story.)

The way the Y-12 process accomplished this separation was simple in principle, though difficult in execution.

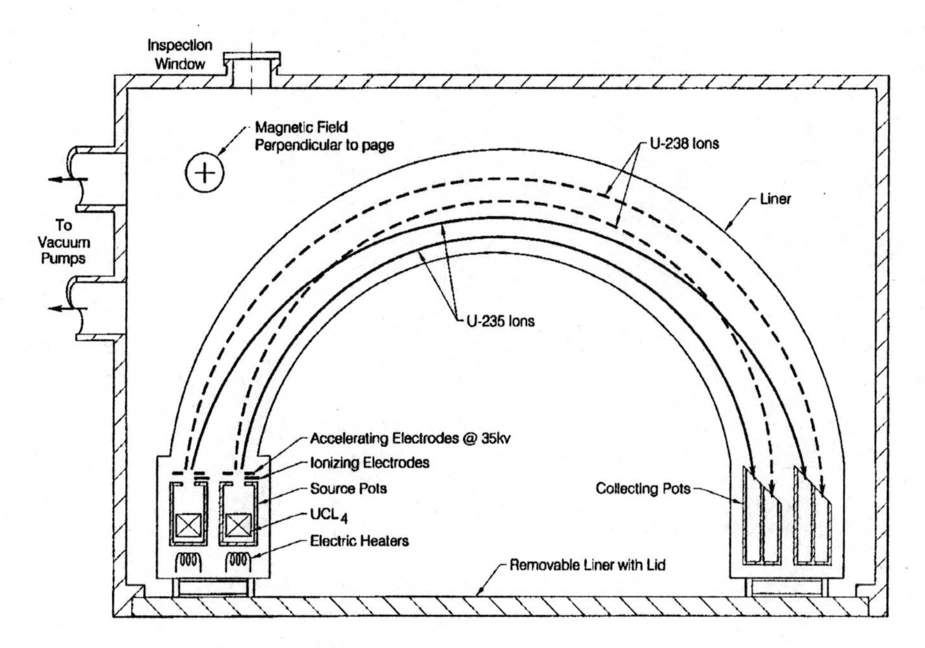

Figure 2.8 Simplified diagram of the Y-12 separation unit for enriching uranium in the fissionable isotope (*J. Nevin Hoke*)

To carry out the process, the Y-12 facility used nine buildings housing 864 first-stage separation units in nine "racetracks" and 288 second stage units in 6 racetracks. In addition, the plant had facilities for chemical processing, research, engineering, maintenance, planning, management, and other auxiliary functions. A chemical salt of uranium

(UCl_4) that can be made to vaporize at a reasonable temperature (within the range of a kitchen oven) is heated within an extremely good vacuum (all but the last one hundred-millionth of the air is pumped out).

Figure 2.9 The uranium enrichment facility at Y-12. One of 15 "race tracks" (*DOE*)

As the vapor escapes through a slit at the top of the heating chamber, a wire filament, heated to a red glow, boils off electrons. These are attracted to an electrically charged plate at the other side of the chamber. The moving electrons ionize the uranium; that is, they knock off some electrons, leaving the atom electrically charged. These charged uranium atoms are then attracted by an electrode and accelerated through the vacuum chamber. But the moving charged uranium ions are in a strong magnetic field, which curves their path into a semi-circle. The U-235, being lighter than the U-238, swings in a tighter circle and finds a collector box at the end of its path. The U-238, like a heavier ball on an elastic string, swings on a wider arc and misses the box. That's the way the separation process works in theory.

Figure 2.10 Process control cubicles for one race track (*DOE*)

Unfortunately, most of the uranium vapor ends up sprayed all over the insides of the equipment. Some of the U-238 sprays into the U-235 box as voltage, temperature, magnetic field strength and other variables wobble a bit off the perfect settings. So the separation is far from complete, and the material collected in the box, now slightly enriched in U-235 but nowhere near pure, is used to feed another pass at the process. It was the job of the Chemistry Buildings to wash the "gunk" off the walls of the equipment, to separately dissolve the material in the collector box, and to convert it all into UCl_4 for another pass.

The U-235 is extremely valuable, and so extraordinary measures were taken to recover it. Sometimes you would see people in white uniforms, down on their hands and knees with a Geiger counter, looking for a tiny speck that might have dropped on the floor. Before they were sent to the laundry, the uniforms were treated with chemicals to recover any uranium particles that might have fallen on them.

My first assignment was to follow through on an idea, previously suggested by one of the managers, to cut down on losses of liquid nitrogen that was put into small stainless steel bottles called vapor traps, to freeze out any traces of water vapor or other volatiles in the high-vacuum systems. Liquid nitrogen, an exotic substance made from air, boils at 320°F below zero and was shipped in tonnage lots into Oak Ridge from Boston in insulated railroad tank cars. It was pumped around the plant through a mile or more of fourteen-inch pipe enclosing special evacuated insulation around a smaller pipe through which the liquid flowed. At these temperatures, normal insulating techniques don't work. Ordinary insulation is designed to provide numerous air pockets. But liquid nitrogen will liquefy air, which then boils away again, a process that very effectively transfers heat, thwarting the insulating effect. So the insulation must be evacuated, and then radiation of the heat becomes the major mechanism for heat loss. A cheap but highly effective insulation was developed for liquid nitrogen; it consisted of an evacuated space, filled with up to fifteen layers of shiny aluminum foil (to reflect the radiant heat), and spaced with expanded paper. Expanded paper is a commercially available product with many short slits in it, which can be stretched out to form a lacy paper spacer.

I was confident that I could save a lot of liquid nitrogen, code-named 714 because its atomic number is 7 and its atomic weight is 14. (Other chemicals were coded differently; some had completely arbitrary numbers). With the self-assurance of my 21 years, I wrote in one of my technical reports:

> Since several million dollars are spent annually on 714 and since only half of the material paid for reaches the [vapor] trap (and only 0.08% of this serves any useful purpose ... an investigation of the losses of 714 and possible remedial action seem in order....A saving of several thousand dollars a day

should be possible without any risk to production and with no interference with operational procedures.

Backing up this claim was a straightforward process, of the sort any new engineering graduate could tackle with confidence. Under the nurturing eye of my boss Oran Miller, who gave me just the right mix of freedom and guidance, I calculated the heat loss (and thus the boil-off of liquid nitrogen) during storage in the tank cars and during the transfer from the tank car to the storage tank. The study showed that the greatest potential savings would come from reducing loss from the traps during operation. The effective function of the trap was accomplished during the first few hours of a run, when the vacuum was being established and the volatiles trapped out of the air. After that, for the two or three weeks of the run, the liquid nitrogen was no longer needed, but there was no way to remove it without releasing the trapped volatiles back into the vacuum. There it sat, absorbing heat from the hot walls around it and boiling away.

My job was to test out a device that provided reflective insulation that could be closed around the trap after the vacuum stabilized. These tests demonstrated that the system worked remarkably well and, with certain other changes, we managed to reduce overall nitrogen loss by 75 to 80 percent. In addition, the shielded trap offered a new advantage: We now had a means of distinguishing an air leak from a water leak. If a small water leak occurred during operation, the trap could be opened and the water vapor frozen out. If it were an air leak rather than a water leak, then opening the shielding had no effect. I felt good about this. It was a small thing, but in my first few months out of school, I had made a difference.

After a couple of such assignments, I was told in September 1944 that Cliff Graham, a former senior engineer at Allis-Chalmers Manufacturing Company in Milwaukee, Wisconsin, was forming a "Tiger Team," formally known as the Process Improvement Team. Seven young hot shots were to cruise around the plant looking for problems to solve or opportunities for improvement, and I had been asked to join. What a Dream Job! I accepted on the spot and never regretted it.

Cliff and his wife, Bonny, were a wonderful support team. They hosted informal parties for the group, and we grew to be very close. The

Tiger Team proved to be a remarkable group of people, and I learned from each of them. Cliff never seemed to supervise; we all had the feeling that we'd been given a completely free hand. He was, however, always there, giving advice and teaching, and of course, he was often the one who came up with things we should investigate. We had access to all parts of the plant, we worked night and day, and we did some wonderful things. But the most exciting part for me was getting into the chemical processing facilities. I had never been allowed in those buildings, and I was anxious to explore. After all, I was supposed to be a chemical engineer.

Figure 2.11 "The Hit Squad."
The Process Improvement Team at play (*Author*)

Chemistry and Chemical Engineering

One of the first people I met in the Chemistry Building was Brooklyn-born Jack Adolphus Kyger. Jack had a B.S. degree in chemistry from Yale and a Ph.D. in physical chemistry from MIT, but he still talked like a regular guy from Brooklyn. He loved convertibles and liked to tell and listen to jokes. He laughed a lot when exchanging stories about adventures in days gone by. He had previously worked in the uranium metal production facility that Mallinckrodt Chemical Company ran in St. Louis, and he had some great stories to tell about his experience there.

"Man, it was hot there! There was no air conditioning in those days, and we used to wrap ourselves in a wet sheet and sit in front of a fan at night to keep cool. The plant produced uranium in hundred-pound metal ingots. Now, you know that uranium is nearly twice as heavy as lead, and lead is half-again heavier than iron. So these hundred-pound uranium ingots were only a tad bigger than a two-quart can. But they still weighed a hundred pounds. We had a great big guy who inspected them and wrote down the ingot numbers that were stamped on the ends. Sometimes the stamped end was down, and this guy would reach over and pick up the ingot with one hand. He complained to me one day that his wrist hurt from doing that. He figured he was getting weak."

Jack's special knowledge of uranium chemistry and chemical engineering processes were much prized at Y-12. He knew his way around chemical equipment in general and the plant in particular. We hit it off from the start, and I soon looked to him as a mentor. The Y-12 process equipment was a showcase for the latest and best that technology could offer: scores of glass-lined reaction vessels and glass piping, and other equipment made of stainless steel, monel and Hastelloy. These materials were considered corrosion-resistant, but each had its weakness. Glass would resist everything but fluorides; stainless, everything but chlorides; monel, everything but nitric acid; and Hastelloy was almost impossible to weld or cut with the technology of the day. The chemical process at Y-12 was continually changing as knowledge grew; at various times we did have to work with fluorides, chlorides and nitric acid, so we always had some equipment we couldn't use.

There were automatic process control systems, all dazzlingly new and shiny. I was awed, but Kyger's reaction was different. "It took us two weeks just to get a batch of pure water through the plant," he told me.

"The automatic valves and level controllers kept shuffling the stuff around so that nobody knew where it was. We finally figured out that we had about a cupful in each of scores of vessels."

"In the Mallinckrodt plant," Jack related, "everything was simple—crude, actually. Each reaction was carried out in a separate room, with a drain in the floor. Workmen walked around in big rubber boots, and didn't worry if some uranium solution splashed onto the floor. Every night the whole room, including the boots, was hosed down and the solution washed down the drain into a redwood recovery tank. The tank was stirred with a wooden paddle. No stainless steel in the whole plant. It was crude, but we had a very high recovery rate. Probably better than this place, with all its nooks and crannies for stuff to hide out in."

Jack's love of the simple in chemical processing equipment showed up in many ways. A key step in the process involved filtering uranium oxide out of a solution. The filtration had to be done just a few degrees above freezing temperature; otherwise the oxide formed a thin slime, which clogged the filter. In keeping with the plant's overall design philosophy, filtration was done with very fancy rotating vacuum filters called "Oliver filters" and centrifuges, with elaborate provisions for maintaining the right temperature. This didn't work very well. Operators spent most of their time shutting down the equipment and cleaning off the slime.

Kyger got the idea of using old-fashioned filter presses—somewhat like apple presses for making cider, except here you threw away the juice and kept the pulp. A few simple calculations showed that a few filter presses, with a block of ice on top of each to keep them cool, could replace two floors of state-of-the-art automatic equipment that wasn't working very well. Jack scrounged up a couple of stainless steel, flat-frame filter presses, and we got permission to work all night testing them out. They worked like a charm. Instead of small quantities of slime that had to be reprocessed, we produced a batch of dry uranium oxide filter cakes, looking like nice, clean Celotex soundproofing tiles (without the holes), and we stacked them up on the Plant Manager's desk. After a proper interval of arguing and retesting, filter presses replaced most of the other filtration equipment, leaving lots of empty space in the building.

Today, the idea of piling stacks of pure uranium oxide on an office desk would shock many people, but it was not considered out-of-line in those days. We did not think of uranium as a hazardous material to handle, as indeed it was not. As one of the scientists, a spectroscopist named Herb Pomerance, told a reporter after the war: "You can't hurt yourself handling uranium—even enriched uranium—unless you drop it on your toe." There were contradictions though. One of the men in our group had an old orange ceramic Fiestaware coffee mug that he brought in to work. (We liked to have our own mugs.) This mug got its color from a common glaze containing uranium. The mug set off the radiation alarms at the gates, and the guards made him take the cup home. It did not meet our requirements for radiation safety. This was also true of some wristwatches containing radium for illuminating the dials. These could not be brought into the work areas either; they were too radioactive. Nobody claimed they were hazardous, but they set off the alarms.

Neutrons, however, were a different matter. Radiation detectors had been around since x rays were first put in use at the beginning of the 20th century. These radiation detectors, however, were designed to read x rays or gamma rays, which are electromagnetic waves, like visible light but with much higher energy. Gamma detectors were not capable of responding to neutrons, and gamma shielding did not necessarily protect people from neutrons. In the early days, there were plenty of portable gamma detectors around, but few survey meters for neutrons. So one of the physicists was surprised when a "health physicist" came around with one of the new portable neutron monitors and told him he had a hot spot on his desk. It turned out that there was a neutron source in the next room encased in a heavy lead shield against the wall. That would have been fine, except that lead is almost transparent to neutrons. They had to get paraffin, which is nearly transparent to gammas, to complete the shield. This was the type of thing we learned early in the game.

Although these radiation fields were high compared with permissible levels, there was such a large margin between *permissible* and *harmful*, that there were no health problems created. Most poisons can be detected only when they get to dangerous levels, but radioactivity is detectable almost at the single atom level, far below the danger point. That's why any radiation detector—even a very cheap and simple one—

will click away merrily, recording the natural background radiation, even when there are no man-made sources present.

The only events that caused significant increases in our background radiation at Y-12 were occasional bursts of fallout from the A-bomb testing in Nevada. "The atoms are coming home to roost," we kidded. In March 1953, we had a particularly high count from the test series, two thousand miles away, where I had been only four thousand yards from the detonation. Not only could we tell that there had been a test, but with careful radiochemical analysis of the fallout to detect each atomic isotope, experts could give a detailed description of the materials used in the bomb, and tell a great deal about the design of the bomb itself. It gave me a new and personal perspective on the sensitivity of our radiation detection equipment.

Post-War Oak Ridge

By the end of the war, the town of Oak Ridge had peaked at 75,000 inhabitants, with a workforce of over 80,000. A few months later, plant construction was nearly over, and Y-12 was laying off thousands of operators. Even so, the residential parts of the town were booming. The citizens reacted variously to this situation. Some planted hedges and flowers and made improvements to their rented homes, confident that some day they would be able to buy them. When no one was looking, they might even dig up and transplant bulbs from houses that had been vacated. Other people adopted a wait-and-see attitude and let the weeds grow. Still others decided to build permanent houses just outside the fence. This town, built as a temporary wartime place, was settling down into permanent existence. The local newsletter had grown from a single mimeographed sheet to a sixteen-page formal letterpress newspaper. You could go to an outdoor theater, a night baseball game, a Tom Thumb golf course, a riding academy, a pony ring or a roller-skating rink. There were already seven movie theaters, any number of dance floors, a drama and an opera group, and an excellent symphony orchestra. Meatless Tuesdays, which had been started earlier in the year, were soon eliminated. But there were still signs advertising "Tent for Rent." A very special town, never dull.

The Navy had been smarter than the Army in the treatment of recruits. The Army had drafted young scientists, engineers and technicians, and made privates and PFCs out of nearly all of them, whether they were Ph.D.s or lab technicians. They lived in crude barracks at the edge of town, with few privileges, as typical "unlisted men" in any army camp. The Navy took the same sort of draftees and made them officers. The Army was thus required to treat these naval officers with good quarters, an officers' club (with the only beer in town), access to car and driver, and other perks befitting an "officer and a gentleman." Not surprisingly, this difference was not lost on the army recruits. Of course, those of us who managed to get there as deferred civilians had the best deal of all.

Figure 2.12 Picturesque boardwalk made from scrap winds through the Oak Ridge woods (*DOE*)

Post-war Lifestyle

With many of the senior scientists leaving, and thousands of employees being laid off at Y-12, the housing crunch began to ease. One day, Frank Ward, who headed one of the reactor mechanical design groups at X-10, called. He had somehow gotten hold of a "D" House, which he could keep if he housed five other men in it. He had one spot left and asked if I wanted to join them. I was delighted to be included. The house had two beds in each of the three bedrooms, a bathroom with a tub, a dining room separate from the living room, a modern, well-equipped kitchen, and a large porch. Frank topped it off with a real touch of luxury: For three dollars a week—fifty cents apiece—we could get a "colored girl" to come in afternoons to wash the breakfast dishes, make the beds and do a little "light dustin'," prepare and serve dinner, and clean up the dinner dishes, too. This was really living!

We were quite proud of ourselves, but we were not the first to have a group house arrangement. I had discovered a house with six women nearly a year before, while housing was still tight. I don't know how they worked it, but they had the same sort of set-up we had (without the maid service). Three of the women we saw frequently, and we still keep in touch with them fifty years later. One, Anne Bishop, was a physicist and worked initially at the Y-12 Pilot Plant when I was there. Anne had a lot of sparkle, laughed a lot, and took obvious delight in all the amazing things that were going on. One day after the war, Anne and I were eating a fast food lunch at work and listening to the news on the radio. The announcer described the work going on with rockets and quoted someone as saying someday we would be able to fly to the moon, and that the whole journey, round-trip, might take less than an hour. Anne's reaction was remarkably down-to-earth: "Gee, it will be faster to get to Knoxville via the moon than by the regular bus." After she left Oak Ridge, Anne went on to get an M.D. degree and conduct arthritis research at Johns Hopkins, in addition marrying the head physician, Victor McKusick, and raising a fine family.

Isabelle Devenish, a tall, impish woman with long blond hair, was perhaps the most outgoing. She had a great Irish mother back in Cincinnati, who took some of us on an all-night tour of the speakeasies across the river in Covington, Kentucky. Isabelle worked in Y-12 as a Job Analyst, and with her brother, Bob, who got there a few months earlier, apparently had enough clout to arrange the housing deal. In April 1945, she left Oak Ridge to join the American Red Cross with the troops in Europe, inspired in part by letters I showed her from my sister Paisley, who was already there. The third member of the household was Jean Benham, who worked in Personnel at Y-12. Jean was one of the few people at that time who wore bluejeans much of the time, and she had the kind of beauty that made you think, *Boy! Would she be a knockout in formal gear!* But in fact, her particular kind of beauty showed off best in bluejeans. All three of the women were college graduates and were smart and fun to be with. There were three others in the house, but those are the three I particularly remember.

One of our favorite activities after the War was to go out for the day on a cabin cruiser owned by six engineers and kept on nearby Lake Norris, behind TVA's Norris Dam. The owners had bought it used for a hundred dollars each, fixed it up, and were generous in the loan of it. There were three or four of us men who used to call on the three women from "the house," and we took great delight in not indicating who was supposed to be paired up with whom. It was all of us dating all of them. Early on, one of the boat owners, Lynn McCabe, brought along Mary Compton, an executive secretary who lived with her sister's family and worked for the superintendent of Lynn's building.

Chuck McVey, one of Lynn's co-workers, and I solemnly agreed that she was "Lynn's girl," and we were not going to muscle in on our buddy's territory, although we also thought she was pretty special. But that evening, being a thoughtful guy, I decided to drop in on her and just see how she had enjoyed the day's activities. While I was there, the phone rang, and when I answered it, who should it be but Chuck. Of course, I roundly reproached him for so quickly going back on his word. For a moment, he was embarrassed, ashamed, and sheepish, and then he caught himself, "Why you old rascal, what are *you* doing there? I'm coming right over!" And so we had another party. In due course, Mary and I got married in Oak Ridge's Chapel-on-the-Hill, but we still keep in touch with Chuck and Lynn. [I know this all sounds like "Leave it to Beaver," but this was the late 1940s after all, and that's the way it was.]

By the end of the war, I was beginning to feel like an old-timer. I enjoyed regaling newcomers with stories about how Oak Ridge was when I arrived on the scene. But occasionally I ran into someone who was here some months before I was, when there was only one building, and the operating companies had not even begun to arrive. In those days, the workers were brought in by station wagons and driven out to the nearest town for lunch. There was not yet any water piped into the area. They passed a water jug around the office from time to time for the workers. These people would end their stories by smiling tolerantly and remarking sweetly that I was lucky to arrive after the mud was tamed and civilization had been founded. I protested mildly at this, but wished fervently that I, too, had been in on the *very* beginning.

Transferring to The Lab

In the spring of 1946, an invitation went out from General Leslie R. Groves, the Army Engineer Corps officer in charge, not only of Colonel Nichols' Manhattan District but also the political and military aspects in Washington. General Groves invited industry and the armed services to send people to Oak Ridge to learn the new technology that had been kept secret even from those with top secret military clearances. On August 1, 1946, President Truman signed into effect the Atomic Energy Act of 1946, taking control of the atom away from the Army and vesting it in a new civilian Atomic Energy Commission. The promise seemed to be that the industrial miracle that created an atomic bomb industry from scratch in less than three years was now about to "harness the atom for the homes and farms and factories of America" and quickly, too. The message was: If you wanted to get in on the action, you'd better send somebody down to Oak Ridge and find out how to do it.

The Y-12 electromagnetic process for enriching uranium was being replaced by the cheaper K-25 gaseous diffusion process, and Kyger and I decided to transfer to the Oak Ridge National Laboratory, or ORNL, as the X-10 site was then called. This was where nuclear reactors were going to be developed, and this, we decided, was where the action was going to be.

A Look Back

When a project works out all right, it's hard, years later, to recapture the feeling that it might have been a fiasco. In the summer of

1944, there was no guarantee that an atomic bomb could be made, especially in the timeframe necessary to end the war. The electromagnetic plant for producing bomb-grade uranium was months behind schedule and could not keep its equipment in sustained operation. The alternative process, the gaseous diffusion plant, was only half built and its basic component, a diffusion barrier that could separate the fissionable uranium isotope from the non-fissionable one, had still not been developed and produced in quantity. The plutonium-generating reactors at Hanford, Washington had not been able to develop a satisfactory way to protect the uranium from corroding and to prevent it from distorting under irradiation. And the weapons laboratory at Los Alamos kept getting thrown off balance by new data on the nuclear and physical properties of plutonium that changed the numbers in their calculations.

General Groves decided that to get sufficient uranium of the required quality, he would have to add additional separation facilities to both the gaseous diffusion and the electromagnetic separation plants. He wrote: "On the assumption that the war with Japan will not be over before July 1946, it is planned to proceed…." As it happened, performance and reliability of the existing plants improved to the point where enough uranium and enough plutonium were produced by the end of July 1945 to make one bomb of each. These were dropped on Japan August 6th and 9th, ending the war a few days later.

But what if the war had ceased a few months earlier, without the bomb? Or what if the problems at Y-12 and Hanford had not been solved in those next few months? There would have been a public outcry over the secret expenditure of $2 billion, an unprecedented sum in those days, and the project would probably have been summarily shut down, with no solid evidence that the release of atomic energy was achievable. How then would our history read? How long would it have taken before we would have had to resurrect the old Manhattan Project and frantically try to find a limitless energy source for an energy-starved planet?

I am convinced that energy will continue to be more and more important as we face an increasingly crowded planet with limited resources. I continue to be awed by the circumstances that allowed me to be in Oak Ridge at that very special time in history. In that context, the handful of students who came to Tennessee at General Groves' invitation was a modest beginning indeed for the loudly trumpeted Atomic Age. But a special curriculum was quickly developed for the new students, and senior scientists already there were allowed to monitor a few courses that they felt would help them in their work. I decided to sit in on some of the courses. I had already taken some pretty exotic stuff from a number of high-powered instructors as part of my earlier training, so I expected to have little trouble with these courses.

That's when I found out that the Navy had also sent a few officers down for this training. They didn't wear uniforms, and there was one silver-haired guy named Rickover who kept asking simple, basic questions. He was obviously determined not to let anything get by him, and unlike the rest of us, he was not at all concerned about making himself look ignorant. I admired him for that.

Science vs. Engineering

Although some awesome structures were constructed at Oak Ridge, involving high-vacuum systems and high-voltage equipment of unprecedented size, and chemical processing facilities dealing with previously unknown materials, there was little engineering going on, in the classical sense. Nobelist E. O. Lawrence, inventor of the cyclotron, and his crew of physicists were developing the devices to separate the fissionable isotope of uranium, U-235, from the much more abundant non-fissionable U-238. They did this by combining their knowledge of theoretical physics with their mechanical ingenuity and an intensive regimen of try, test, and adjust. They built the hardware by adapting cyclotron design to the new requirements of uranium isotope separation. When they finally had some-thing that seemed to work, Lawrence cried "Ship it!" and it was sent to Westinghouse where hundreds of identical copies were made. Never mind that as it was patched together, some of the copper tubing was 5/16-inch, some ¼-inch, and some 3/16. *We know this particular design works; it took a lot of finagling to get it there; don't change a thing!*

This was a perfectly proper procedure for a crash wartime project, and the technical intuition and ingenuity of the scientists who did it earned my undying admiration and respect. But it was not engineering. In the next couple of years, I was to learn in painful detail just how big a difference that made.

♦ ♦ ♦ ♦

In September 1992, I was invited to attend the 50-year anniversary celebration of the town of Oak Ridge, and to address the local Rotary Club on the contributions of Admiral Rickover to nuclear power. How the town had changed! I had not been there for 25 years, and I could hardly recognize any of the landmarks I associated with Oak Ridge. There was one exception: the Chapel-on-the-Hill, where Mary and I had been married 45 years before.

I discovered that the unchanged appearance of the Chapel was no accident. The Town Council had formally decided long ago to leave untouched the site of so many baptisms, confirmations, weddings and funerals. Other than that, the town looked like so many other modern, bustling, American cities, with enclosed shopping malls, divided highways, and some rather pretentious, privately built "rich folks' mansions." I was glad to see that much of the original "temporary" housing was still standing, having been sold to private owners. Most of it had been made almost unrecognizable by the addition of extra rooms, brick or stone facing and extensive landscaping. The basic structures, inside and out, had held up remarkably well.

3. Getting the Atom Away from the Army

The Two-Edged Atomic Sword

The anti-nuclear movement is generally viewed as coming from the political left, and thus nuclear power advocates are often assumed to be protecting the military-industrial status quo. This strikes me as bitterly ironic as I recall the massive grassroots effort we nuclear scientists and engineers mounted fifty years ago to wrest control of the atom from the Army and "give it to The People." It's a story that would warm the heart of Jane Fonda.

In 1945, when the reality of the war's end finally sank in and the celebrations died down, we were left with a feeling of optimism and confidence. I was in Germany in 1989 when the Berlin Wall was coming down. The great Brandenburg Gate was being opened, and the pulse of the people there was much the same as I experienced in Oak Ridge in August 1945. In both instances there was a feeling that the forces of oppression had been overcome, the People had triumphed, and we were about to enter a new era where common sense, good will, and the voice of the individual citizen would prevail. We had finally learned to overcome war and tyranny, and we were going to see that they never again established a foothold. At Oak Ridge, we felt we couldn't count on the characters who had been running the world up to now, and we decided we were going to have to do something about it ourselves. Why not? Hadn't we ended the war in a matter of days? Hadn't we harnessed the mightiest force in the universe? We were ready for the next chore. With the brash arrogance of youth, we gave little thought or credit to the people who had envisioned and managed the mighty effort that created the Bomb; we saw only the technical efforts of the working level scientists and engineers. To us, the rest was merely office work.

The papers and the radio were full of talk about the implications of the fissioned atom for the future, how important it was to protect "the secret," to develop a defense … and we knew they were talking nonsense. There was no secret a spy could steal that would enable a foreign power to build an A-bomb. We knew that any industrialized nation willing to put the effort and the resources into it could learn to build a Bomb, no matter how hard we tried to protect "the secret." The most important secret had been revealed at Hiroshima: it could be done. The Smyth Report described how. There was little else to protect.

As for defenses, we were concerned that while the U.S. developed more and more sophisticated anti-missile devices, a determined enemy could smuggle in bombs in moving vans or even small delivery trucks. If we continued to build bombs, we would only provide a greater incentive for another nation to attack us before we got more. The only answer, we were convinced, was to get the Bomb and atomic energy research and development away from the Army, and give it to The People—conceive some civilian agency working out in the open, where it could be monitored and controlled by an international agency that had the confidence of all the people on the planet. What a wonderful world we envisioned! We summed it all up in the slogan: "NO SECRET. NO DEFENSE. THEREFORE: INTERNATIONAL CONTROL."

The Quixotic Lobbyists

At every site associated with the Manhattan Project, and then with other interested parties outside the project, scientists and engineers spontaneously got together and talked earnestly about what they could do to steer national and international events in the right direction. They formed groups, the most active and effective of which were at Chicago (where the very first nuclear chain reaction was demonstrated) and, to a lesser extent, Los Alamos, New Mexico (where the bomb design and fabrication was carried out). Even in those days of limited communication, this rapidly became a nationwide phenomenon. Although my part in the unfolding events was minor, I was keenly involved and was in a good position to see what was going on and who the key players were.

Very late one night in my dormitory room, a few weeks after the war's end, a few of us were going over, for the umpteenth time, how we, in our naiveté and isolation could influence this global situation. In addition to me, there were two physicists, one chemist, a chemical engineer, and a student from Princeton's famed School of Public and International Affairs. Average age 23, all were employees of the electro-magnetic separation plant at Y-12.

"This United Nations Organization isn't even officially in business yet, and people are already trying to kill it," said Dave Wehmeyer, 22-year old physicist from Detroit. "I think we'd better support it, or we may never get another international organization." "Wavey Davey" had started work a few weeks before I did, so he was my first boss at the plant.

"From what I've read, it's got some serious weaknesses," said Jack Balderston, 23-year old chemical engineer. "It may not be able to do the job."

"Then I suggest we send for a copy of the UNO Charter and find out what's wrong with it," said Dieter Gruen, chemist, in his faintly European accent. (Foreign accents were not unusual in Oak Ridge; many of the leading scientists were fugitives from Hitler's Europe. A congressman once complained, "Aren't there any scientists with American names?")

Bill McLean, 23-year old chemist, burst in impatiently: "Aw, let's just write to the key guys in this thing—say the President's Interim Committee, Vannevar Bush, 'Satchel-ass' Groves, and the rest of them—tell 'em what we want to do, and ask 'em how to go about it. The straightforward approach. That'll confuse 'em."

Of course, at that point we weren't yet clear ourselves on what we wanted to do. We did get a copy of the draft United Nations Charter, and after many hours of heated debate, we developed a list of amendments we felt were needed. We sent these off to the Interim Committee and a few others and received various non-committal responses.

Some weeks later, Dr. Harry Pearlman, brilliant young MIT-trained chemical engineer who had joined the group asked, "How do we get this from a bull-session to some sort of political action?"

We agreed that we had to get more people into it, to somehow develop a grassroots movement. But then Pearlman raised a critical question: "Do we state our ideas and then sign up those who agree with us, or do we pull in everyone who claims to be interested in peace, and see what ideas come out?"

That really started a donnybrook. One side argued, "We've battled this thing out, several hours a day for nearly a month, and we all agree that world government and international civilian control of atomic energy are the only rational answers. I can't see abandoning that carefully arrived-at conclusion to the first rabble-rousing nationalist we sign up, nor can I see spending a month convincing each new member of the wisdom of this conclusion. We need to speak with one clear voice."

The other side responded, "Who are we to tell a group of hundreds—maybe thousands—what to think? Are we the only thinkers in Y-12? I say sign 'em up and see what comes out. That's the only way everyone will be behind this."

The question was settled in an unexpected and exciting way. It turned out that most of the technical people in the plant had already been thinking and talking together about these things. They had individually embraced as foregone conclusions the ideas we considered too radical for ready acceptance. When our statement of intent was presented at the first mass meeting, it was quickly passed unanimously. A line of thought held by less than a quarter of the American population at large was the spontaneous unanimous opinion of these atomic scientists and engineers!

Meanwhile, at X-10

Dr. Joseph H. Rush, who had been a physics professor at Denison University in Ohio and was very active in the post-war action, recalls how this spontaneous mobilization process started at X-10, where he worked. Clinton Laboratories, as it was then called, owned the facility doing pilot plant and development work in support of the production and separation of plutonium, the alternative approach to atomic fission. "We were all annoyed at the announcement after Hiroshima that there would be no holiday in the event of an armistice. We were told to continue regular wartime work hours until VJ Day, the formal surrender ceremonies. But by that time, nobody was interested in celebrating."

In the July 2, 1960 *Saturday Review,* Joe Rush recalled that day as follows:

> On the day of the armistice, nearly everyone in Clinton Laboratories came to the plant as ordered. In the Physics Division, we drifted into the library and began to talk. Little conversational nuclei took shape, men sitting on chairs or tables or just standing. The driving purpose that had ordered our energies was gone, and I think everyone felt to some extent a sense of disorientation, of slackness, of loss of direction.
>
> The evolution of that day's discussion was remarkable. Certainly everyone had given some thought to the long-range consequences of the bomb and the problems it would raise after the war.... Yet on this day of armistice we did not talk immediately of these larger issues. We griped about the denial of a holiday, and the poor food in the plant cafeteria, and the inadequate bus service. As these common irritations were aired, the little knots of conversation melted and flowed into a more general participation, and the discussion began to find direction. It was as if we had to recapitulate consciously the frustrations and vexations that had been denied outlet, to bring ourselves up to date emotionally, before we could look into the uncertain future.
>
> Even then, our first concerns were for our own professional prospects, and for the future of Clinton Laboratories and other atomic enterprises ... Especially we wondered what role the military would play in postwar atomic developments. We knew as did few others that the bomb represented not merely a weapon but a radical new technology, and we felt strongly that atomic energy and the problems it would create needed to be dealt with through open, democratic processes. Near the end of that holiday in the physics library, we found ourselves confronting gingerly and with only rudimentary awareness the key questions that were to engage us so intensely in the times ahead. People would need

to be informed, educated to the potentialities of this new frontier. They would have to be warned of its terrible threat, assured of its hidden promise.

Trying to Convey the Message

This unanimity of feeling among ourselves, and a great suspicion on the part of most of the public that yielding to international control was somehow "giving in to the Russians," made for a lot of letters home and letters to editors. As an example of this dialogue, I quote from a letter I wrote to my father on September 30, 1945:

> Everyone will soon be arguing about sharing or not sharing the "secret" of atomic energy and I thought you might be interested in hearing my two cents worth. In the first place, there is no secret that we can hide or share, any more than there is a secret of how to make Fords which no country ever duplicated. It is really not possible to keep the secret. If we attempt to do so, it will mean stifling information here and there, greatly slowing down development on the biggest field since fire was discovered....
>
> I don't see how there could ever be a defense against the bomb. What I mean by that is that the bomb can be smuggled into any city by agents, and the best plane detectors in the world would be useless.
>
> We all feel here that the only course is completely unrestricted publication; even if some countries don't play fair, we still come out ahead...

As the months passed, the tone grew more emotional. I wrote home on July 1, 1946:

> If half the effort that is being put into plans to make the next war last days instead of hours were put into eliminating it altogether, it could be done. World government carries no implications of bowing down to Russia or anybody else; it merely means that you are tired of being a sucker for treaties and are determined to set up a government with enforceable laws. Is that idealistic? Is it less realistic to say that we are not ready for world brotherly love, where a treaty means something, than it is to try to establish a two-fisted government with the purpose of maintaining law and order?

> Does it make you happy to know that the Army is planning to disburse and bury cities and make more atom bombs, so that after our cities are wiped out we can wipe out everybody else's, and that dispersal will make the destruction of our cities take days instead of minutes? Thus we "win" the war. That is more realistic than preventing?! Who wants to prevent it? That's leftwing talk.
>
> Pardon the soap-boxing, but I get pretty disgusted sometimes with the way that people refuse to face the obvious.

The proposals we drew up were sufficiently detailed and sufficiently different that they led to a lot of reading, research and discussion. They also led us to some nationally known speakers who could guide us in further town meeting discussions. We argued a lot and learned a lot, but the result of all this was mostly an increase in our own knowledge and sophistication in matters political. We did not produce any startling new political insights or proposals.

So we argued with the folks at home, and we argued with the people in Washington. I've been asked how the local Tennesseans felt about the issues. My answer is: I really don't know. One effect of compartmentalization and long workdays was that we didn't have much political or technical discussion with people outside the circle of our professional colleagues. The red-fearing chauvinism often attributed to the rural south may have contributed to the hostility that sometimes surfaced between Ridgers and Locals, but we tended to attribute it mostly to an insensitive elitism we often unthinkingly projected. Even today, when a cultural event such as a concert by a world-class musician or entertainer takes place in Oak Ridge, the Knoxville papers generally ignore it.

The Snowball Grows

This compartmentalization also made it difficult to reach beyond the Y-12 group to include more of the atomic community. We didn't even *know* people who worked at other sites. Even the adult education classes in physics, chemistry, math, and the like, were segregated by companies. The obvious way to reach others was through public announcements, but such announcements would normally be cleared through the Army's censorship and public relations people. It would be nice to get their blessing, but if they turned us down, we didn't want to stop there. Would we be better off to go around them?

This question, too, was answered simply and unexpectedly. In October, the Knoxville *News-Sentinel* carried a two-column story on page one, reporting the existence of a group calling themselves the Association of Oak Ridge Scientists at Clinton Laboratories. This group had written a Statement of Intent that read as if it had been written by *our* group. The similarity was astonishing. This event taught us that one could publish such an article without getting Army clearance. Luckily, the Army Public Relations Officer, genial Lieutenant George O. "Gus" Robinson, Jr. took no retaliatory action, other than rubbing his balding head with the heel of his left hand and wrinkling his tired brow. We learned later that he was working on a book of his own about life in Oak Ridge.

The X-10 group was way ahead of us. They had already signed up 96 percent of the scientific personnel at their lab and had contacted important personages in Washington. Their conclusions about the crisis and their approach to doing something about it were so nearly identical to our own that we felt a great boost to our morale. The third largest Oak Ridge installation, the gaseous diffusion plant at the K-25 site, soon announced its own organization with similar purposes, the Atomic Engineers of Oak Ridge. This group actually got Colonel Parsons, the chief security officer, to address them about the Army's attitude toward such groups. This action went a long way to clearing away some of our fears and misgivings.

Two tasks now faced us, both of which were less fun than discoursing on world politics. First was the tedious chore of drawing up by-laws, appointing committees, electing officers, and all the other bureaucratic chores that technical people usually dodge. Second was the need to consolidate the various Oak Ridge groups, each of which was used to acting as if it was *the* spokesman for atomic scientists. The Y-12 group and the K-25 group merged rather easily to form ORES (Oak Ridge Engineers and Scientists). The X-10 group, the Association of Oak Ridge Scientists at Clinton Laboratories had dropped the laboratory name at the request of management and become AORS (Association of Oak Ridge Scientists). They had polled their group extensively and knew they had near-unanimity on the key issues. Understandably, they were concerned about joining our group, which was something of an unknown to them. It was already clear, if ironic, that the supposedly ivory-towered physicists were studying the particulars of specific legislative proposals, with the intent of influencing them in detail, while the pragmatic engineers tended toward educating themselves on various long-range proposals for international control, and even world government.

By November 1945, only three months after Hiroshima, ORES had 600 members, the X-10 group had 170, Los Alamos 300, and the Chicago group 200. Scientists at other war projects had sprung up, bringing the total nationally to about 3,000.

The League of Frightened Men

The question of consolidation was again unexpectedly resolved, this time by a phone call from Washington. Dr. John A. Simpson, a 29-year old physicist from the Manhattan Project's Metallurgical Laboratory in Chicago, was calling to say that the Chicago group had been caught off-guard by the sudden introduction of the Army's May-Johnson bill in the Senate. This bill would establish a commission outside the Army to carry on all atomic work, but the Army would still exercise considerable control over all atomic research and development. The hearings were forced through in one day, less than a week after its introduction, and the House Military Affairs Committee was already meeting in executive session to consider it. It might soon be law.

We were stunned. None of us was familiar with the details of the bill and the rumors we had heard about it bothered us. A month before, the Chicago group had held a public meeting in which the brash young physicist Samuel Allison quipped that if the Army insisted on continuing its onerous security restrictions, its scientists might all go off and study butterflies. This resulted in a sharp rebuke from Colonel Nichols, head of the Army's Manhattan District, who said that such talk might interfere with the administration's legislative proposals. The scientists replied that no one had informed us about the specifics of these proposals, and we were, therefore, not in a position to be concerned about that.

Simpson went on to urge that it was now clear we all had to upgrade both the magnitude and the effectiveness of our effort. To that end, he and others were setting up a Federation of Atomic Scientists (FAS), with which all the other groups could be affiliated. He was a bachelor and figured that his meager personal funds would support him for about a year, long enough to get the organization started. The FAS would be the eyes, ears and voice in Washington of all the atomic scientists and engineers.

A tiny office on the top floor of the building at 1016 Vermont Avenue was quickly set up with a phone, some desks piled high with hand-outs, speeches, and news clips, a file cabinet, a mimeograph and a secretary. Scientists from the other groups could drop in any time, be briefed as to status and urgently needed action, and would be sent off to proselyte policymakers and bring back notes of interviews for the office records. Score was kept by listing key players as "scared" or "unscared;" the purpose of the visits being to convert them from the second category to the first. These quixotic crusaders became known among more sophisticated Washington operatives as The Quiz Kids, The Friends of the Atom, The Reluctant Lobby, or The League of Frightened Men—sobriquets that did not hurt their image or their cause.

The Press found them a novelty. In a city where lobbyists were nearly always hired guns, speaking for whomever happened to be their client at the moment, these people were speaking for themselves. They were the genuine article, a primary source. Yet they weren't fighting for something for themselves; they were fighting to save the world. Beirne Lay of *Life* described the operation as "a test-tube of unadulterated democracy.

The organization had no president or chairman, because nobody wanted to be czar. The members came to Washington, not to get something, but to give something—to give the most precious commodity in existence: knowledge." Historian Alice Kimball Smith, dean emerita of Radcliffe, wrote in *A Peril and a Hope: The Scientists' Movement in America, 1945-47*:

> "Without salary, without a publicity director, without political know-how, without staff or office equipment, without Pullman reservations, and without arrogance, they had come,

> bringing knowledge, sincerity, patience, humility, and a desire to perform a public service."

Visitors marveled at how junior scientists would argue as equals with Nobel laureates on any question that came up for discussion. This was the way of science, but it was rare in politics. Later, when the organization had moved to a fifth-floor walk-up at 1621 K Street, between two Chinese restaurants, Mike Amrine, a savvy reporter came aboard to help with public relations. He wrote: "As they told the world what the bomb could do to civilization, I saw what the bomb had done to these professors." But when one of the members in their endless discussions suggested that the world would be better off if scientists were in charge, Amrine pounded his fist on the table and yelled, "I'd rather be bombed!"

The various atomic groups made use of the Washington office, but the matter didn't stop there. The bold agenda of the atomic scientists attracted other socially active groups outside the atomic fraternity, including women's groups, labor and religious organizations and others of various stripe. There were 49 such groups with a constituency of over ten million members, but the FAS was extremely leery of affiliation of any kind.

The FAS leadership felt their effectiveness depended on a perception of political naiveté and purity that could be tarnished by association with more experienced political groups with particular agendas and historical baggage of their own. To maintain that position and avoid any appearance of being a tool for any other group or agenda, the FAS turned down a potentially lucrative offer from MGM Studio to use the FAS name as technical advisor for a motion picture account of their work. They returned a $5,000 advance check from MGM. That was serious money in those days (more than a year's salary for most of us), and the decision took considerable moral courage. Similarly, the organization rejected another offer from a radio producer to accept a retainer to supply information on an exclusive basis.

The FAS handled relationship with other interested parties adroitly by setting up a National Committee on Atomic Information (NCAI), which it controlled. The NCAI put out newsletters and information kits on atomic energy but took no stands on political issues. By keeping a loose connection with the other groups, FAS was able to some extent to have it both ways. To accommodate the interests of other scientists, with whom the members felt more at ease on political matters, FAS agreed to establishment of a Federation of American Scientists. Under the able leadership of William A. Higinbotham of the Association of Los Alamos Scientists; Melba Phillips, Secretary; Joseph H. Rush, Treasurer; and with reporter Mike Amrine as Publications Editor, this "other FAS" worked closely with, and finally supplanted the Federation of Atomic Scientists.

These were not your typical office clerks. Willy Higinbotham was a Ph.D. physicist, widely admired for his innovative designs of nuclear instrumentation and the inventor of "Pong," the first video game (in 1958!). Dr. Melba Phillips had been a Fellow at the famed Institute for Advanced Study at Princeton, and later was fired from Brooklyn College and Columbia Radiation Laboratory for refusing to name names for the McCarran Committee. Joe Rush had a Ph.D. in physics from Duke, and spent most of the rest of his life at the National Center for Atmospheric Research, from which he was called to assist the Condon Committee with preparation of the Blue Book report on Unidentified Flying Objects. His primary scholarly interest in those years was in "exploring the boundaries of human capability," the subtitle of his monumental *Foundations of Parapsychology.* Mike Amrine was a freelance investigative reporter who was one of the first to publicize the Navy's unwillingness to promote Captain Rickover, publicity that led to Rickover's ending his 63-year naval career 30 years later as a four-star admiral.

A book—or many books—could be written about the political actions of the next few months. Led by the Chicago group, an intensive educational program was set up to inform important decision-makers about the facts of nuclear energy. (We could start with how to pronounce it properly!) Trips were arranged to Oak Ridge and Los Alamos, and informal technical seminars were hastily put together. An ambitious freshman senator from Connecticut, Brien McMahon, seemed most receptive, and the young scientists enjoyed playing teacher for such illustrious students. We were awed by the politicos, but Jack Kyger commented to me that he was surprised how impressed, and even deferential, many of the congressmen were with regard to *us*. It was a new experience for *all of* us.

Earlier Efforts

Even before the end of the war, although we in Oak Ridge were unaware of it, some of the top-level scientists on the project had already been lobbying on their own, and not always toward the same end. As early as January 1944, the eminent physicist Leo Szilard wrote to Vannevar Bush, the President's science advisor, urging that work on the bomb be expedited. He argued that unless full-sized atomic bombs were actually used in the war, the public would not understand the magnitude of their destructive power and would not be willing to pay the price of peace.

Then on June 12, 1945, seven Chicago scientists led by Professor James Franck delivered a memo to Bush with quite a different message: that a demonstration detonation of the bomb should be given for UN officials at some remote, uninhabited location, prior to any military use. Shortly thereafter, a similar recommendation, signed by 64 scientists at the laboratory, was sent directly to President Truman. Truman gave the Franck proposal to a panel of four eminent scientists: Arthur H. Compton, Ernest Lawrence, J. Robert Oppenheimer, and Enrico Fermi. After anguished consideration, the four scientists unanimously concluded that direct military use, without warning, seemed to be the only feasible option. Navy Undersecretary Ralph A. Bard disagreed, arguing for further attempts to negotiate with the Japanese. But Secretary of War Stimson concurred with the panel, and Truman accepted this recommendation.

Even among the Chicago group, the scientists were not all of like mind. A multiple-choice poll by Compton of 150 project scientists taken shortly before Hiroshima showed that nearly half favored "a military demonstration in Japan." (It is not clear whether the respondents would have considered Hiroshima, which was an Army headquarters site, a port of embarkation, and a convoy assembly point, as well as a manufacturing center, to be in this category.) About a quarter of the respondents preferred "an experimental demonstration," and 15 percent chose "use in whatever manner the military believed would end the war with the least loss of American lives." Eleven percent asked for a public demonstration but no military use, and 2 percent asked that the technology be kept "as secret as possible."

In our naiveté we didn't even consider the extensive time and planning required for any of these operations. Unbeknownst to us, the personnel and special equipment required for the A-bomb runs over Japan had already been selected more than a year before. The military had long been in training to carry and drop the large and clumsy weapon and to execute the special evasive maneuver developed to get the aircraft out of harm's way after the bomb was released.

We Post Our Theses

Trying to put our own views into words that were both clear and rational on the one hand, yet sufficiently passionate and persuasive to arouse others to action, was a new type of challenge for us. In August 1946, as spokesman for the Federation of American Scientists, I wrote a piece called "Credo of an American Engineer" for *This Week*, the national Sunday newspaper supplement. The article was accepted for later publication but never actually printed. In it I listed a number of brief statements, each followed by a paragraph or two of amplification, summarizing the principles that guided our political action groups. Excerpts from this credo follow:

1. *I believe there can be no order without law, no law without government, and that this is as true on a world scale as it is for your city.*

2. *I believe that a treaty between nations is as uselessly idealistic without world government as a written agreement to stop crime in your city would be without city government. Law must reach the individual.*

3. *I believe a peace enforced by an alliance of two or three strong nations will last about as long as the "thousand year peace" of the Berlin-Rome-Tokyo Axis.*

4. *I believe disarmament and appeasement alone will prevent World War III about as well as it prevented World War II.*

5. *I believe nuclear energy is as fundamental as fire, and that it cannot be kept secret or controlled by the military.*

6. *I do not believe that the military is as capable of handling science as are scientists.*

7. *I believe that when we are "ready" for world government, we will no longer need it; this "unreadiness" is the surest sign of our crying need for it.*

The question of whether we should push for world government, as argued above, or call merely for "international control" of the atom, was always a bone of contention. The engineers tended to believe that only world government would work, whereas the scientists tended to focus on less radical goals. The FAS encouraged the site groups to study and discuss various long-range plans for international control, but it steadfastly refused to comment publicly on any proposals for partial or complete world government such as Harold Stassen's, Ely Culbertson's or Clarence Streit's.

Inspection and Detection

A key factor in evaluating any mechanism for control was the question of inspection and verification: What steps could an inspection agency take to ensure that material was not being diverted clandestinely for military purposes? This question occurred to the congressmen early in their deliberations, and scientists and engineers could help in addressing it. At the request of Congress, a number of detailed technical reports were prepared describing how an agency might carry out an inspection and auditing operation of ostensibly peaceful atomic facilities to detect illicit diversion. These reports were indeed helpful in clarifying what could and could not be accomplished in monitoring a non-proliferation agreement. Unfortunately, they were highly classified and thus, not available to the public or to anyone else outside the small circle of people authorized to read such reports.

As late as 1965, I was asked to co-author a report that was carried out with access to top-secret documents but was finally published as a hard-cover unclassified book, *Arms Control Agreements: Designs for Verification and Organization*, (D.W. Wainwright, *et al* Johns Hopkins Press, 1968). We concluded that a modest number of trained technicians, sampling various process streams in the plant and using customized statistical procedures and accounting concepts of comparing numbers, which should have known relationships, could probably do an acceptable job of detecting any significant diversion of fissionable material. The large numbers of senior scientists called for in some of the other verification studies did not seem necessary to us. Our book never received much public attention, but I was asked to make a draft of the section on nuclear facilities available on short notice for a breakfast meeting between President Johnson and Soviet officials. I have no idea whether it had any impact, or whether it was even used, but it made me feel good at the time.

We Win One

The first significant victory for the scientists' lobby was getting the House hearings on the Army's May-Johnson bill reopened for a second day, but it was a victory short-lived. What we in the various scientist/engineer groups were after was, first, that atomic energy be seen as more than a weapon. Second, that the security measures that would impose severe penalties for vaguely defined offenses be loosened and clarified. Finally, no policies or actions should be implemented domestically that would impede efforts to ultimately internationalize control over atomic weaponry. We thought there was virtually unanimous agreement among us on those points, which had been stated in numerous proclamations. And we were convinced that the May-Johnson bill worked against these goals. We were about to learn one of our first political lessons.

Some of the top scientists of the project testified but, surprisingly, were little help. Leo Szilard's testimony was seen as rambling and unfocused. Herbert Anderson came across as hostile and dogmatic. Arthur Compton and J. Robert Oppenheimer testified that the May-Johnson bill, which we were fighting, was acceptable. Harold Urey was to testify last, but the hour was late and he could not be found. Chairman May remarked, "The War Department discovered the weapon. Why can they not keep the secret?" and closed the hearings.

Figure 3.1 U.S. soldiers destroying Japanese cyclotrons as "war making materials" (*National Archives*)

Our optimism surged and ebbed as events both onstage and off alternatively brightened and then dimmed our hopes. In November 1945, U.S. Army engineers and ordnance men with cutting torches and demolition charges raided research laboratories in Osaka, Kyoto, and Tokyo. They destroyed five cyclotrons and all experimental data obtained with those devices, under Army orders to eliminate anything that could contribute to Japan's war-making potential. In response to public uproar, General Groves admitted that this was a serious blunder, and the scientists played it up as an example of the Army's inability to understand scientific matters. Three months later, the Canadian atomic spy case broke just as Congress was debating how stringent to make the security requirements protecting atomic energy information. The Army and their congressional allies used this episode to strengthen their case.

Somehow, despite all these distractions, things kept moving in the Congress. Senator McMahon had introduced a bill to create a special Senate committee on atomic energy, and on October 23 that committee was created. McMahon was appointed chairman—quite a prize for a freshman senator—but his influence was tempered considerably by the conservatism of the other ten appointees to the committee. After further fieldwork, President Truman was persuaded to privately withdraw his support for the May-Johnson bill, leaving it up to others to create an alternative proposal. By the end of the year, McMahon was ready to introduce his own bill, and on June 1, 1946, McMahon's bill was passed. Truman signed it into law as the Atomic Energy Act of 1946.

We had scored a significant victory. The atom was to come under a fully civilian agency, the U.S. Atomic Energy Commission. The new law was designed to emphasize research and the development of peaceful uses of atomic energy; provide "free dissemination of basic scientific information;" "maximum liberality in dissemination of related technical information;" and "Government control of the production, ownership and use of fissionable materials." It was clearly a good launching pad for working toward international control.

On June 14, 1946, Bernard Baruch presented to the opening session of the United Nations Atomic Energy Commission the American proposal for controlling the atom:

> We are here to make a choice between the quick and the dead.
>
> That is our business.
>
> Behind the black portent of the new atomic age lies a hope which, seized upon with faith, can work our salvation. If we fail, then we have damned every man to be the slave of Fear. Let us not deceive ourselves: We must elect World Peace or World Destruction...
>
> Science, which gave us this dread power, shows that it *can* be made a giant help to humanity, but science does *not* show us how to prevent its baleful use. So we have been appointed to obviate that peril by finding a meeting of the minds and the hearts of our peoples. Only in the will of mankind lies the answer.

The scientists were pleased with much of the proposal, but many felt Baruch had sabotaged the attempt to find accord with the Soviet Union. And then came Bikini. Since October 1945, the Army and the Navy had been talking about running a test of the bomb against naval vessels. These discussions quickly became a replay of the parochial arguments and issues raised by Billy Mitchell's demonstration of air power against naval vessels after World War I. The scientists were concerned about many aspects of these tests, carried out near the Bikini atoll in the far Pacific. We feared that America's credibility would be damaged, by urging peace and restraint on others while we carried out military demonstrations of negligible scientific value. There was also concern that little attention was being given to ensure radiological safety for the participants. But foremost, we knew that the bomb would probably not directly sink many of the heavily-armored warships, spread out over miles of ocean, and the public would suddenly conclude that the bomb's destructive power had been overblown.

Just two weeks after Baruch's stirring challenge, the Army dropped the first bomb at Bikini. Gabriel Heatter's soothing voice assured radio listeners: "The palm trees are still standing on Bikini tonight." And the respected *New York Times* writer William L. Laurence wrote of "the profound change in the public attitude" caused by the demonstration:

> Before Bikini the world stood in awe of this new cosmic force.... Since Bikini this feeling of awe has largely evaporated.

And so the national and the international politicians fell back into familiar channels, and few bold new measures were undertaken. Nonetheless, in America, a civilian Atomic Energy Commission began business, and in Vienna, an International Atomic Energy Agency was ultimately brought into being (1957). At this writing, bureaucracy as usual seems to be the order of the day, but the atomic warfare we all feared has not yet broken out.

Trying to Reach the Russians

One of the frustrations that burned in the gullets of the young scientists, naive as we were concerning the art of diplomacy and politics, was the inability to sit down and talk with the Russians and "work this thing out." *How tough could it be, really?* we thought. The Russians had been devastated by the war, and we knew they were not anxious for another. Bertrand Russell, the maverick British philosopher, joined with Albert Einstein and other prominent scientists from various countries to discuss how scientists could help in cooling the arms race and ameliorating some of humanity's other pressing problems. They managed to involve a few scientists from Russia and its allies, but there was no indication that government policies were being affected.

Talking with friends in the intelligence community, we learned that a major problem was the Russian mode of communication to high officials. Anything originating in America that was considered of possible interest to higher-ups was translated by bureaucrats who were anxious to demonstrate their allegiance to strict Leninist principles. So if an American official put out a feeler, the first translator would report it from a purely a Marxist perspective. If it were deemed important enough to pass up the chain of command, it would be condensed and given further spin. Consequently, the message that finally reached a Russian policymaker was always that the Americans were offering yet another capitalist trap.

We discovered that there was a way around this game. The McGraw-Hill Book Company, as it was then known, published a series of engineering trade journals, *Mechanical Engineering, Electrical Engineering*, and the like. These journals ran editorials that sometimes ventured into the realm of national policy as it affected industry and technology. The Russians were interested in learning everything they could about American technology, and these magazines were translated by relatively apolitical technologists and selectively read by policymakers. There was thus, a channel outside the normal diplomatic morass where one could float trial balloons. We had some indications that these were in fact reaching their target, although we could attribute no political breakthroughs to this activity.

Shifting Into Educational Mode

We had wrested the atom away from the Army and given it to the People (so we thought), but we had made little progress toward our goal of promoting international control. The next step, we decided, was one of public education—a big job and a necessarily slow one. The National Committee on Atomic Information, which the FAS had set up, was largely a firefighting outfit, rebutting charges of communism, straightening out gross errors of fact, and supplying basic information on nuclear science. For the broader issues we set up an Association of Scientists for Atomic Education (ASAE). We divided the country into regions, and local chapters of the ASAE were established under various regional councils. Naturally, close working relations were maintained with the FAS.

Illustrating the depth and specificity of ASAE's intentions, the following were the topics suggested by the Board of Directors to each Region for discussion and preparation of regional resolutions. Each of these proposals, which we evaluated in lengthy discussions, studies and correspondence, was a particular plan for achieving an enforceable world peace:

1. The Szilard "Call for a Crusade."
2. Urey's "Alternate Course for Control of Atomic Energy."
3. Osborne's Popular Convention to Frame a Constitution for a Federal World Government.
4. The Montreux Declaration on World Government.
5. The Squires-Daniels-Cavers proposal for a moratorium on atomic production.
6. The Gromyko proposal on atomic energy control.
7. The British Association of Scientific Workers' proposal for atomic energy control.
8. The Marshall Plan (European Recovery Program).
9. The Szilard "Letter to Stalin" proposals.

The memorandum was signed by Jack Balderston, who had participated in that early discussion in my dormitory a year before.

In addition to the work of scientists and engineers to study and to educate themselves in these matters, a series of town meetings led by noted speakers was held in the high school auditorium, the only such facility in town. These were stereotypical American town meetings in that each citizen who wished to comment on the subject at hand was given respectful attention and time to do so. The meetings were unique, however, in the global sweep of the issues covered. They were memorable affairs, and each had its own tone and power. Ely Culbertson, for example, surprised most of us by saying he had spent much of his early life in foreign jails as a political prisoner, and that he had devoted much of that time to studying possible forms of world government. He said he had created the card game "bridge" as a bet with a psychologist friend that he could invent a game that would sweep the world. To him it was an experiment in mass psychology, nothing more, but now it no longer occupied his mind. When asked long and rambling questions, he would repeat them verbatim, then paraphrase them into crisply worded questions, pause a moment, and answer with similar brisk clarity. He had a detailed plan of action, and specific answers to every question. It was a dazzling demonstration of a powerful mind at work, and the entire hall was entranced.

The next meeting featured the noted writer and editor Norman Cousins, a totally different phenomenon. My main recollection from that meeting was the emotional intensity that he built up, in stark contrast to the Culbertson meeting. I remember a woman stepping out into the aisle and walking toward him, her arms outstretched, tears running down her cheeks, sobbing, "But what can *I* do, Mr. Cousins? What can *I* do?" He replied, with equal fervor, "Shout it in the streets! Knock on doors! Storm the Capitol!"

Charles D. Coryell, a radiochemist from X-10 and a student of Glenn Seaborg, gave a talk to the high school students, and they were sufficiently moved to organize the Youth Council on the Atomic Crisis (known as "Yak-Ack" among the irreverent). In short order they managed to get themselves heard over national radio, had articles in the national press, and were invited by the UN Council of Philadelphia to address groups there with a total audience estimated at 21,000.

These and other political actions were effective. When the House tried to load the McMahon bill with onerous amendments, 70,000 letters of protest were received at a time in history when public participation in the political process was otherwise at a low ebb. And the process continued for another decade. The *Bulletin of the Atomic Scientists* continues publication to this day as a widely read journal of opinion and information. However, I share the disappointment of Alvin Weinberg, former director of Oak Ridge National Laboratory, who wrote in *The First Nuclear Era* (AIP Press, 1994):

> As so often happens with such organizations, FAS and the *Bulletin* were gradually captured by anti-nuclear activists.... I am particularly chagrined that the *Bulletin*, which under its first editor, Eugene Rabinowitch, saw nuclear energy as a powerful agent for creating material abundance, now seems to view nuclear energy as an abomination.

One of the lessons we learned from lobbying was that the most effective motivator was a message of impending doom. We were willing to use this tactic to get people's attention in the effort to achieve civilian control of the atomic bomb. But we were quite unprepared for the same tactic to be used against nuclear power in the 1970s and beyond. Perhaps we had it coming to us.

REDS IN OUR ATOM BOMB PLANTS

The Full, Documented Story

By Representative J. PARNELL THOMAS

A congressman who has dug out the facts warns that our atomic-energy secrets may be secrets no longer

REDS in our Atom-Bomb plants

By Representative

J. PARNELL THOMAS

as told to STACY V. JONES

READING TIME • 14 MINUTES 40 SECONDS

[EDITOR'S NOTE: Representative Thomas is chairman of the House Committee of Un-American Activities and has served on the committee since its formation in 1938. He is also a member of the Armed Forces Committee and, until the new House rules limited committee memberships, served on the Joint Atomic Energy Committee.]

THE Atomic Energy Commission must come to grips shortly with pro-Soviet infiltration of its own organization. Fellow travelers, if not actual members of the Communist Party, have, for instance, ensconced themselves in the great plants at Oak Ridge, Tennessee, where U-235 is separated for use in the atomic bomb.

On a recent visit to Oak Ridge, I was startled to find how many Communist suspects were on duty there—in the electromagnetic plant, the gaseous-diffusion plant, or the Clinton Laboratories where general nuclear research is conducted. The laboratories are the most heavily infested.

A Communist policy enforces concealment of party membership on persons holding responsible government jobs, and there can be no doubt that, in addition to the known adherents, many others are on the payroll. I examined Army Intelligence reports on a number of men holding strategic positions. Several of these dossiers showed, in my opinion, very serious cases.

A natural question is why the known followers of the party line haven't been fired. The commanding officer assured me that the matter was very delicate, for if certain of the suspected physicists were discharged, scores of other scientists had threatened to walk out. This inability to throw out persons of doubtful loyalty is a serious matter.

Communist penetration has ap-

(Continued on page 90)

JUNE 21, 1947 LIBERTY 15

Figure 4.1 J. Parnell Thomas launches red scare at Oak Ridge (*Liberty*)

4. Fighting the Red Hunters

There is another part of the early atomic history that many of today's anti-nuclear activists have conveniently forgotten, or perhaps never knew. Starting right after the war, the scientists and engineers, who had unleashed this new force, found themselves under vicious and personally threatening attack from government groups and their allies crusading against suspected communists. These groups considered any suggestion of supplanting Army control of atomic research to be un-American. In our efforts to establish international civilian control of atomic energy development, we opposed the Army's initial proposal to maintain control. Ultimately, our position was adopted by the majority of Congress and was signed into law by the President. But even after that, many of us came under personal attack as Communist dupes or agents.

The Un-American Activities Committee

One of the earliest, most vociferous and most persistent of the attackers was Congressman J. Parnell Thomas (D) of New Jersey, Chairman of the notorious House Un-American Activities Committee. On June 4, 1946, he sent two investigators down to Oak Ridge, who claimed to be sympathetically interested in the aims and the programs of the Oak Ridge scientist groups. The investigators were freely shown through the files and reports and invited to a meeting scheduled for that evening. But they left after four hours. On July 11, Thomas threw a bombshell into the committee hearings in the form of a report by Ernie Adamson, the committee's chief counsel, claiming "a six month investigation," of serious security problems at Oak Ridge. The report charged that some scientists who used to work at Oak Ridge continued to correspond with scientists "inside the reservation," and charged that groups had been formed that were "definitely opposed to Army supervision at Oak Ridge." The report went on to say. "The security officers at Oak Ridge think that the peace and security of the United States is definitely in danger." As if charges of treason were not enough, the report also charged the scientists with working with the CIO and the AFL to unionize the plants.

The scientists, speaking through the Association of Oak Ridge Scientists (AORS), as well as for themselves individually, hotly blasted the report and responded to each of the charges. They noted that all of their activities had been in the open and under the watchful eye of Army security personnel. Furthermore, their positions were consistent with those expressed by the Secretary of the Army, the majority of Congress, and the President of the United States. Oak Ridge security officers were questioned by the committee and flatly denied having expressed any concern for the national safety. The scientists noted that they had voluntarily agreed to stop publishing their nuclear research in 1939, and had "kept the secret" for three years before the Army created the Manhattan Project. And they were completely mystified as to the charge of working with the unions. And whatever their personal feelings about unions, they were just not in a position to get into that struggle. By showering officials and news media with letters and factual statements, they unequivocally demolished the Adamson report. But that was not to be the end of it.

"They Call It Security"

A year later, Rep. Thomas wrote an article entitled, "Reds In Our Atom Bomb Plants: The Full, Documented Story," which ran in the June 21, 1947 issue of *Liberty*, a popular weekly magazine of the day. The story, "as told to" a professional writer, was consistent in tone with "The Woman with a Scar" and "Washington Murder Go Round" in the same issue. The article was an amplification of the same kind of charges made in Adamson's report. A frightening red hammer and sickle was imposed on an aerial photo of Oak Ridge (Figure 4.1). What Thomas probably didn't know is that the fearsome emblem was centered directly on the building with the largest product output in Oak Ridge: the hospital's maternity ward.

Although the Atomic Energy Act of 1946 had long been the law of the land, Thomas concluded, "I believe that in the present chaotic world situation our only solution is to repeal the act and return Manhattan District to the Army, which can best administer security. Again, the scientist groups sprang into action, giving interviews, writing letters, and refuting the various allegations. I was really annoyed that widely-publicized charges as specific as this could be made by a person in such an authoritative position, and never proved or disproved. I was determined to do something about it, or at least raise hell! I wrote an article for *The Saturday Evening Post* that had the following note under the title: "This article was written at the request of the Association of Oak Ridge Engineers and Scientists by one of its members and was reviewed, discussed, revised, and approved by them before publication." The article was entitled "They Call It Security," and noted that Webster's dictionary linked security with being "free from fear, care, or anxiety; easy in mind," but that Webster had never been to Oak Ridge. Then I got to the meat of the matter:

> Representative J. Parnell Thomas recently charged that 'our atomic energy secrets may be secrets no longer,' that U.S. atomic scientists are 'fellow travelers, if not actual members of the Communist Party,' and that 'if certain of the suspected physicists were discharged, scores of other scientists had threatened to walk out.'... If Mr. Thomas knows of any such agents, he should do as was done [in Canada]: gather his information secretly, report his findings to the correct government agency, have them accuse a named list of men with a specific list of crimes, and bring the suspects to trial.... Since there is every indication, strengthened by observation of previous attacks by his committee, that he has no intention of following his overall smear with specific accusations, a reply by the accused is demanded.

I then commenced to respond to each of the charges in turn. First, that his committee presumably had the most extensive files on suspected communists and the responsibility to take whatever action was indicated. Second, with regard to the alleged threat of scientists walking out if certain suspected communists were discharged, I provided a copy of a letter from the head of the Physics Division of the X-10 lab, stating "I can assure you that there is no foundation whatsoever for such an accusation." I went on to refute each of the other charges, also pointing out that it is hard to disprove such charges as "pro-Soviet infiltration...fellow travelers ... communist suspects ... there can be no doubt that many others are on the payroll...persons of doubtful loyalty ..." etc. I also noted that all of the scientists, *alleged communists included*, who had worked there prior to the recent civilian takeover, had been investigated by the Army. Having veto power over all hires, any red infiltration occurred under Army surveillance.

After finally getting approval from the AORES, I showed the article to the AEC public information people and promptly received a request to see one of the Atomic Energy Commissioners. At Oak Ridge, this was like an order from the Pope, and I hurried somewhat nervously to the appointment. The Commissioner told me, in the most conciliatory tones, that the Commission was, at that very moment, trying to build special relations with the Congress, and they would appreciate very much if I held off on the article. They were confident that problems of this sort would soon be straightened out, and I wouldn't want to spoil that, would I? Of course I wouldn't, and the article never saw the light of day. But Thomas wasn't through with us yet.

"They've Taken My Badge!"

One day, one of my fellow workers came up to me with a wild and frightened look. "They've taken my badge," he sobbed, choking back tears "What am I supposed to do? They've taken my badge!"

"What did you do?" I asked. "Why did they take it?"

"Nothing!" he said. "I didn't do anything. I don't know why they took it. They just called me into Security this morning and took my badge away. They didn't say why. They didn't say when they'd give it back. They didn't tell me what to do to get it back. I don't even know if I'm allowed to stay in Oak Ridge. They didn't tell me anything!"

I didn't know what to say to him. I was speechless. A couple of other people walking by had overheard the exchange—he was talking pretty loudly—and someone broke in to say that another young scientist had had the same thing happen. This wasn't an isolated case. The other victim decided to head for the AORES office to get what support he could, which we all agreed seemed like a reasonable move. The various scientist groups immediately recognized the gravity of the security action and its longer-range implications, and they pulled out all available firepower. Stephen White did a piece in the *NY Herald Tribune* headlined TWO ATOMIC SCIENTISTS SUSPENDED, MANY MORE FACE LOYALTY INQUIRY: "Oak Ridge Hearings Based on Anonymous Charges of Red Leanings; Suspects Say Accusations Abound in Errors, Which Can Be Checked."

The article quoted from the official charges. The first defendant had four charges, the first two of which stated: "1. A former landlord of yours has reported that in 1943, after you moved from the premises, certain magazines and pamphlets which may have been left on the premises by you may have included a copy of the magazine *New Masses*. 2. A neighbor has stated that she believes a close relative by marriage is a communist." The other two charges were similar. The second defendant had only one charge, which stated in its entirety: "A person with whom you associated closely in the years 1943–47 said you were very enthusiastic about Russia and seemed to be pro-Russian in your view." Three other scientists were under investigation, with the charges against them similarly vague.

The newspaper columnist pointed out several easily checked errors in the charges and stated, "Similar errors occur in almost all cases." He quoted defendant number two: "Who is this man that says I am a communist? Who am I defending myself against? He has no name, no face, no social security number." Marquis Childs, a nationally syndicated columnist, wrote under the headline: CASE BEFORE LOYALTY BOARD ILLUSTRATES THE PRESENT STATE OF NATIONAL NEAR-HYSTERIA. Tom Stokes wrote another column that was particularly pointed, headlined WHAT WAS THE CRIME OF SCIENTISTS AT OAK RIDGE?

My father wrote me a letter in the midst of all this, enclosing some inflammatory clippings from the *Chicago Tribune*. "This is the sort of thing you have advised us was apt to appear," he wrote. "I think I understand thoroughly your point of view in this situation, but do not become too fanatical over it. Projects of this magnitude take time and patience, and investigators can so easily distort what you say…"

I replied, "…Remember: we are running no 'secret movement'. We are working with Congress and with the press, and they with us. There is nothing shady in what we do. You mustn't believe everything the *Trib* says.… I really have been fairly pleased and proud of the way things have gone. The Congressmen who knew the facts were very much on the right team, and most of them were willing to listen. Notice that the Senate, after intensive campaigning by us, passed the McMahon bill, which was good. Then the House, whom we hadn't had time to work on, murdered it. Then, in committee, when the Senate boys, now on our team, explained the thing, it passed overnight. I think there is still hope for the ole U.S."

J. Parnell Thomas Gets His

Well, the People ultimately did tire of the excessive tactics of the red-baiters, but not until a large number of individuals had lost their jobs and their reputations, and even ten-year old Shirley Temple had been accused of aiding the forces of subversion. Finally, Joe McCarthy was censured by his colleagues, the Un-American Activities Committee was disbanded, and J. Parnell Thomas was thrown into jail. Thomas's end was particularly ironic.

Born John Patrick Feeney at a time when discrimination against "shanty-Irish" was prevalent, he changed his name to evoke the patriotic image of Charles Stewart Parnell, militant Irish nationalist of the previous generation. Not satisfied with a congressman's salary, Thomas began in 1940 to add names to his congressional payroll and pocket their salaries. This went on for eight years, until his secretary, Helen Campbell, who was having an adulterous affair with him, discovered his infidelity and told all to columnist Drew Pearson. Ironically, the career of Charles Parnell, the Irish patriot whose name Thomas borrowed, had ended the same way.

After getting five trial postponements by a variety of means, including going in for unnecessary surgery, he tried for a sixth, but the doctors would not admit him for further treatment. He was sentenced to a federal penitentiary for 18 months and fined $10,000 for embezzling $8,000.

Half-a-century later, newly available Russian records revealed that Congressman Samuel Dickstein (D) from New York, one of the founders of the House UnAmerican Activities Committee, was, in fact, a paid agent of the KGB, sending periodic reports to Russian intelligence while denouncing fellow Americans as being "soft on communism."

Getting Beyond the Bomb

With the Bomb apparently safely in civilian hands, and an International Atomic Energy Agency struggling to be born, our attention began to turn toward broader issues. How could this awesome force be utilized for peaceful purposes? There was talk of using atomic explosions to dig ditches and move mountains and even to power spaceships. There was a proposal to blast a sea-level canal through Panama, firing off nuclear explosives in a carefully choreographed sequence, to peel back the earth like opening a zipper. But environmentalists raised serious questions about the consequences of directly connecting the two oceans, and that plan was dropped at an early stage.

A new breed of scientists, engineers and technical managers was pouring into Oak Ridge, eager to explore a vast rainbow of possible atomic reactors—*piles* they were called in those days, after the original practice of piling up uranium and graphite blocks. The combinations of fuels, coolants, structural materials, and moderators, feeding a wide variety of power conversion systems—steam turbines, gas turbines, and direct conversion of electricity from flowing hot ionized gases—offered a nearly limitless field for research and development. Oak Ridge would have much to keep it occupied in the days and years ahead. Harnessing the "Beast" would be a worthy challenge; no young engineer could help feeling a thrill at the chance to be one of the "few, we happy few, we band of brothers" (and a few sisters), privileged to undertake this important task for humankind.

5. From Science to Technology

June 1946 started off like every other June in Oak Ridge. The Great Smoky Mountains were at their loveliest. spring was rapidly blending into summer; the dogwood, rhododendron and mountain laurel were in full bloom. One could almost see Disney's baby unicorns and centaurs gamboling through the perfumed meadows. One thing was new: General Groves had invited industry and the military to send people down to study this new atomic technology via a one-year curriculum hastily put together by the project scientists, and the new "students" were beginning to arrive. The lecture program arranged for them had lots of ground-breaking material, information that had not yet appeared in any textbooks. It was clear that the year would be absorbing and challenging for them.

They were a heterogeneous assemblage: self-assured managers, on hand to get a quick overview then return home to start up a nuclear division; eager scientists and engineers, anxious to get a technological jump on their peers; and blasé Army officers, for whom this was just another tour of duty. All were there as individuals, but one day a technical report appeared, and then another, and another, all authored by "The Naval Group." Each of these reports summarized the state of knowledge of a single technical aspect of nuclear technology. They were very good.

We were curious about these authorless reports, and checking, we found the trail led to one Captain Hyman Rickover, who was now at Oak Ridge with four other officers and three civilians. Each of them had been told they were there as independent visitors and were to report back individually to their Bureau superiors in Washington. So what was this Naval Group?

How that came about was told to me by Lieutenant James M. Dunford, USN, Dunford had just packed up his family and moved half way around the world for engineering duty with the Pearl Harbor Naval Shipyard. He was well qualified for that job, ranked number three out of 581 in the Naval Academy class of 1939 and had a master's degree in naval construction and engineering from MIT. He and his family were looking forward to a stimulating and pleasant tour of duty in the Hawaiian Islands. But he had been ambitious enough to formally volunteer for "any special or unusual assignment," and so he had been ordered to Oak Ridge. He was planning how best to organize his time there when he received a message from Rickover, asking him to meet with him that evening. He had no idea why the Captain wanted to see him, but the question was quickly answered at the meeting.

In addition to Dunford, Rickover had assembled the other three naval officer students. Eschewing pleasantries, Rickover opened the meeting with a simple statement: "As senior naval officer on the site, I've been given authority for filing your Fitness Reports while you're here. I presume you all understand the various implications of that responsibility. Keep in touch with me. I will be discussing with you how we can use this opportunity for maximum advantage. That will be all for now." And he turned on his heel and walked out of the room.

Dunford and the other officers stayed in their seats, silent for a moment. No one had to spell out that the officer who files your Fitness Reports holds your career in his hands. They didn't know exactly what would come next, but they each knew that their personal game plans had taken a sudden change in course. Most of the other officers were of the same quality as Dunford. Louis Roddis, Ned Beach, and Dunford had competed for top standing in the class of 1939 through four years of the Naval Academy; Roddis ended up number one. Beach, who covered himself with glory as a wartime submariner and later as author of the popular submarine novel *Run Silent, Run Deep*, was number two. (Beach was not at Oak Ridge, but he would later enter the program and sail the USS Triton around the world submerged.) Dunford was number three.

The other career officer, Lieutenant Commander Miles Libbey, class of 1940, became interested in the physics of the game and stayed on at Oak Ridge after the others went to Washington at the end of the session. The fourth officer, Lieutenant (junior grade) Ray Dick, was a different case. He was a metallurgical engineer from Ohio State and had served as an underwater demolition expert with the Navy's frogmen during the war. He had not volunteered for the Oak Ridge assignment, and as a naval reservist, he was due to retire. But because of his exceptional academic training and his exemplary war record, the Navy tapped him for Oak Ridge, and he voluntarily extended his military service time to take advantage of this unique opportunity. He was brash, outspoken, cynical and openly disdainful of the reverence traditionally due to Naval Academy graduates with distinguished family histories of naval service. Despite the fact that they outranked him significantly, he argued bluntly with all of them, including Rickover. He became, however, one of Rickover's most trusted advisors until he died in January 1953, at age 31, of physical exhaustion. He was the only one of the group who turned to alcohol to ease the stress, and this probably added to his medical problems.

They didn't have to wait long to find out what Rickover had in mind. It was clear that sitting through a bunch of lectures was not going to make nuclear engineers out of them. The Oak Ridge staff had done a good job in the short time available to them, converting their new knowledge into teachable form. But it was aimed at a diverse audience, with widely different backgrounds and interests. It necessarily skimmed the top of a very broad technology that included physics, chemistry, metallurgy, heat transfer, fluid flow, mechanical design, thermodynamics, automatic control theory and practice, and some advanced math, in addition to the specifics of nuclear reactor core design. Each of these topics in turn covered several sub-topics: for example, metallurgy of uranium, of steels and stainless steels, of aluminum alloys, etc.

Rickover's first action was to get each of his new team members to pick one of these basic topics and write the definitive report on it.

"Everything the lecturer says, all his handouts and references, anything you can find in the technical literature, anything any of the students knows from his own work—everything anybody ever knew on this subject, I want in a single, well-organized, clearly-understandable report. Got it?" As soon as they finished one report, they started on another. This required them to work like demons, but Rickover apparently felt they had to learn that lesson anyway, and the sooner the better. These reports, signed merely by "The Naval Group," were quickly seen by all to be the real sourcebook of choice on any topic they had covered. When anyone needed to look up something, the first question was: "Has the Naval Group done a report on this?" If they had, the job was much easier.

The unstated implication behind the Oak Ridge program was that people were going to build nuclear reactors, and the material being taught was what they needed to know. There was actually a small group investigating the problems associated with one particular design concept suggested by chemistry professor Farrington Daniels of the University of Wisconsin. But, in fact, no one really expected that anyone would be building a power reactor very soon. Uranium was thought to be a highly limited resource, and it was destined to be made primarily into explosives. So Rickover's intensity and sense of urgency seemed strangely out of place, both in Oak Ridge and in Washington. But he managed to get his small team fired-up, and they kept their enthusiasm and their drive even amidst their less passionate colleagues. .

Rickover was struck by two features of Oak Ridge. First, there were a lot of very smart people there. But, second, the whole program was being treated as a science project, not an engineering development program. He found there was very little engineering in evidence. That would have to change. I was curious to learn where his intuition would lead us. I had an inkling of what this would mean, from my work at Oak Ridge as well as my engineering education. But I had a lot to learn.

The Difference Between Science and Engineering

I now realize that many people do not fully appreciate the difference between science and engineering, and a number of misconceptions arise as a result, since scientists and engineers have very different ideas and purposes. We are told that Science has brought us such wonders as antibiotics, jet airplanes, and DVDs. But, in fact, Science has done no such thing. Science produces only knowledge. That is its purpose and it does it superbly. Scientists study *what is*. Engineers envision and then build *what never was.* Scientists persuaded the atom to release its binding energy in an uncontrolled burst of fury. It was then up to engineers to tame that fury and make it into a docile and dependable servant of humanity.

Research is what scientists do; *technology* is what engineers do. Scientists may spend most of their time tinkering with a complex piece of hardware—an "atom smasher" or a giant telescope—but it is only a means to an end. Once this hardware has achieved its purpose of enabling a greater understanding, it is of no further use to the scientist. Thus, scientists and engineers work in a completely complementary and symbiotic relationship: scientists use hardware to develop knowledge; engineers use knowledge to develop hardware.

Herbert Simon has suggested that whereas scientists study the laws of the natural world, engineers study the laws of the artificial, the things we have built ourselves. The world of wind tunnels, pilot plants, and nuclear reactors is the realm of research engineers. My engineering company, MPR Associates, once worked with astronomers on a major new telescope design. The astronomers were primarily interested in the astronomical information they would be able to get with it, whereas we were interested in how to meet the design objectives of the device itself—extreme physical rigidity and precise placement of the parts. We were both working on the same product, but for entirely different reasons. This difference did not create any problems; in fact, it enabled each of us to do what we did best.

Scientists study some particular natural phenomenon, and then publish a paper suggesting a *theory* on how they think it works—Einstein's theory of relativity, for example. Then someone suggests a way to test the theory. They propose a *hypothesis:* If this theory is true, then such-and-such must follow, and we ought to be able to construct an experiment or a calculation to see if it does in fact act that way. Because scientists have been ingeniously inventive in devising experiments to test theories, we can build up considerable confidence in any theory that survives this process of scrutiny and testing. After a while, if a theory hasn't been shot down through experimentation or further theorizing, it is said to *explain* the phenomenon (although it may only describe, or even just label it). Finally, this theoretical description is called a *law of nature*.

Many people think of engineering as *applied science*, but that is not accurate. That would be like calling physics "applied math," or calling biology "applied physics." Each of these fields—engineering, math, physics, and biology—is a separate discipline, with its own terminology, its own ground-rules and logic and its own set of basic facts and beliefs. Engineering historian Walter Vincenti shows in some detail that engineering has its own source material and its own modes of thought, quite different from those of science.

Engineering often involves *design optimization.* This requires creating a theoretical model of the intended product and then seeing how the product changes as each individual characteristic, or parameter, of the product is varied on paper or by a computer. Consider the automobile. Some parameters of interest are: cost, ruggedness, speed and agility, stylishness, carrying capacity, luxuriousness, etc.

No single car design is best for everything. If we want to carry lots of children and sports equipment, we would emphasize carrying capacity and get an SUV or a van; however, we would have to sacrifice something in cost, speed and agility. Or we could choose to emphasize speed and agility and get a sports car. If cost is not an objective, we can have a luxury sedan, but a limousine is not very good for hauling cinder blocks or two-by-fours.

This way of looking at a thing holistically is an engineering concept, called systems engineering. Ecologists make much of this approach, but they did not invent it. They borrowed it from engineers. If ecologists were to look only at single symptoms (as an unimaginative physician might diagnose) they might want to destroy locusts or beavers.

Only by looking at the whole ecology as a system of interrelated processes can one make intelligent judgments as to what is happening.

Scientists' hardware is used only by themselves, and they can, therefore, be informal in defining its operating procedures and potential hazards. But engineers turn their hardware over to unfamiliar operators, so they must write down in clear, user-friendly detail how the device should be used, protected, maintained and repaired. They must envision all sorts of potential misuse and warn against it—the "Don't-eat-the-daisies approach." Airplane wings have signs THIS IS NOT A STEP. Cars carry the warnings UNLEADED GAS ONLY. Thus, the engineer gets involved in codes and standards, in procedures and manuals, in inventory systems and other paperwork unknown to the scientist.

I want to make clear, although it should go without saying, that neither science nor engineering is superior to the other, intellectually, spiritually, nor in terms of its central importance to the human race. They are like eating and breathing, quite distinct from each other, but equally important to our survival and well-being. I wish I could dispel the viewpoint described to me by the late Professor "Tommy" Thompson, then U.S. Atomic Energy Commissioner, of his experience as head of the Nuclear Engineering Department at MIT. He described for his students the challenging task of designing a modern nuclear particle accelerator with its huge magnets, multi-million-volt high-frequency electrical system, ultra-high vacuum systems and so on, and then asked: "What would you call the people who work on such systems—scientists or engineers?" Without hesitation, they replied, almost in unison, "Scientists." "Why?" he asked, surprised by the unanimity and certainty of the response. "Because scientists work for Truth and engineers work for profit." He was as dismayed as I was.

Rickover and his Team Return to Washington

Rickover finished up his fifteen months at Oak Ridge by taking his team on a whirlwind tour of facilities and people that might have something to teach him about the status or the future of atomic energy. He was disappointed but not surprised to find that no one was seriously planning a nuclear power program. When he returned to Washington in September 1948, the Navy broke up his group and scattered it, and he was told there was to be no nuclear propulsion program. Another captain was in charge of "atomic matters," but this encompassed only atomic explosives and defense. In Rickover's mind, he was still in charge of his old team, and he kept in close touch with them. He would pull Dunford, or Dick, or Roddis, or various combinations of them, into his office and ruminate over the next possible steps.

"The Navy has sole responsibility for designing, building, maintaining, and operating warships," Rickover mused. "Nobody is going to take *that* away from them. But this new Atomic Energy Commission has sole responsibility and authority for nuclear materials, nuclear safety, and all the classified nuclear information, and that's not going to change soon. So there's only one way to proceed: Anybody who wants to build and operate a nuclear warship had better have line responsibility in both agencies."

His conclusion seemed reasonable, but there was no precedent for how to implement it. Rickover took his usual head-down, plow-straight-forward approach and drafted a Memorandum of Understanding, to be signed by the Secretary of the Navy and the Chairman of the Atomic Energy Commission (AEC). The memorandum recommended that WHEREAS the situation just described existed, THEREFORE there would be a Nuclear Power Branch established in the Navy's Bureau of Ships, with such-and-such responsibilities and authority, and the AEC would set up a Naval Reactors Branch within the Reactor Development Division, with responsibilities and authority as defined. Through the MOU, Rickover further indicated that, not only would these two organizations work closely together, they would be composed of largely the same individuals. It took original thinking to create this document and incredible patience and skill to guide it through the many layers of bureaucracy in the Navy and the AEC, rewording countless times to please the various signatories along the way. In time, however, the task was finally accomplished.

Figure 5.1 With nuclear power still only a dream, Captain Rickover and group look over model (*NY Times*)

Such requirements generally originate with the Navy's Ships Characteristics Board, working with the Bureau of Ships. But Rickover's efforts to get such a statement had met with total rejection. He therefore decided to go straight for Admiral Chester Nimitz, the Chief of Naval Operations. "Hell, he's a submariner," realized Rickover. "He'll understand." As predicted, when the memo finally reached him on December 5, 1947, having been pushed up through the ranks for two months, Nimitz listened to the rationale and signed the memo without a change. He also persuaded the Secretary of the Navy to sign the accompanying letter to the Secretary of Defense and to the Atomic Energy Commission. The Navy was finally on record as requiring an atomic-powered submarine.

The Atomic Energy Commission was another matter. The agency itself had been signed into existence on January 1, 1947, but was still nearly without personnel. It took another six months of intense pressure from the Navy before the AEC finally agreed "to formalize" work on a naval propulsion system. Ten months after that, Dr. Lawrence Hafstad, Executive Secretary of the government's Joint Research and Development Board, was appointed and agreed to serve as the first Director of Reactor Development, effective February 15, 1949. The Naval Reactors Branch was to operate in that division.

Rickover was disappointed that an ex-physics professor was to direct what he considered to be an engineering program of immense proportions. But s it worked out, Hafstad proved to be quite supportive and generally gave Rickover the backing and elbow room he needed. It had been a long fight, and there was still little substance behind the organizational facade, but Rickover at last had an official platform to stand on. Now all he had to do was to begin developing the nuclear power technology to build a nuclear submarine—and then build one.

The Oak Ridge Shielding Conference

In Oak Ridge in 1948, I was largely unaware of all this and was busily working on the engineering and materials aspect of radiation shielding design. Various scientists at universities and research labs, as well as the Manhattan Project sites, had been working on different parts of this problem, but because of security compartmentalization, few people knew what was going on at other labs. I decided to arrange a classified conference on the subject, bringing in all the workers in the field, as well as potential users such as Rickover.

Because of the difficulty of exchanging classified information among labs, I required each speaker to bring 250 printed copies of his talk, along with any further supporting information he considered useful. I then had a complete set of papers bound and distributed them to each participant, with copies left over for appropriate classified libraries. The conference lasted three days, and I even persuaded the laboratory to come up with private money for a beer party for participants, which was pretty far-out for those days. I was feeling good about how the conference was going when I got a call to meet with Captain Rickover at the Oak Ridge Guest House (the closest thing we had to a hotel).

"What are you trying to get out of this tea party, Rockwell?" he asked.

While I thought it was obvious, I explained that by bringing the producers of data and theory together with potential users such as himself, a great deal of useful information exchange would naturally take place.

"Yeah, yeah, that's just dandy," he interrupted. "Then they all go home and what happens next?"

I still wasn't getting it, so he laid it out: "Could you sit down right now and design me a shield?"

"No, of course not," I replied.

"Why not? What do you need to know first? Write it down. Do you need experiments? Who can do them? When could they get started? Are the theories adequate? How do you know? Hell, do I have to spell it all out for you? You ought to get each organization to agree explicitly that they will provide certain experimental data or answer certain theoretical questions by specific dates. Users, like myself, should pose questions they need answered and assign priorities to them. I'm ready to do that. But I want your help on it. Can you work tonight?"

Now I got it, all right, and it looked like a lot of hard work. I had a million things that I needed to be doing to keep the conference on track, but he was of course dead right. And this was a perfect opportunity to really achieve something concrete out of what might otherwise become just another jolly talkfest. This was September 1948, and the previous June, Rickover had hired Jack Kyger, my friend and mentor from Oak Ridge. Without waiting for my reply, Rickover said, "Work with Kyger, and show me what you've got before the meeting starts tomorrow. Kyger will show you what *we're* asking for. I want you to come up with what we should get out of each organization here, and by when." And he turned and went to his bedroom to read reports.

In those first meetings, Rickover struck me as the classical paradigm of a wholly rational, linearly logical thinker. I had been educated and trained to believe that there was no other defensible way to think, and Rickover seemed to embody that idea. So I had to repeatedly face instances of Rickover's being "right for the wrong reasons"—of having the right answer but not being able to explain why—before I could fully appreciate Rickover's ability to draw on intuition and trust it. A very contradictory guy!

In November 1949, about a year after the Shielding Conference, Rickover asked a small number of us working on radiation shielding at various laboratories to come to his office for a Saturday meeting. He wanted to get some definitive answers as to what was needed before we could start a serious design effort. He wanted to determine what kind and how much shielding would be needed on the first nuclear submarine. I cancelled my weekend plans and gave a lot of thought as to what I would say. When I got there, I asked him if I could speak first. He asked me why, and I said that if there was agreement on my suggested approach, it might render moot a lot of what other people would otherwise present. He seemed bemused by my request, but let me start the discussion after his introductory remarks.

I felt I was fairly far out and alone on what I was about to present but proceeded somewhat nervously. I said I thought that a program to reduce shield weight by using exotic materials and sophisticated theoretical and experimental studies would not reach fruition for several years, and would be very expensive and entail unforeseeable problems. Moreover, by the time we considered the effects of cooling the shield materials and providing for necessary structure and penetrations, the final shield design would probably not be much lighter for having used those special materials. I suggested we could probably do about as well by using common shielding materials and calculations based on simple bulk measurements at Oak Ridge.

Although it wasn't apparent from reading the stream of reports people were writing on esoteric materials and calculational techniques, I found that some of the other researchers were beginning to reach the same conclusions I had. After some lively discussion, it was agreed that this was the way to go.

As the meeting broke up, the Captain called me back. "How much are you making, Rockwell?" he asked. I told him, and he turned to Dunford: "We can beat that, can't we, Dunford?" As Dunford nodded agreement, Rickover shouted to his secretary: "Get me Al Weinberg!" (Weinberg was the Research Director of the Oak Ridge laboratory.)

> "Wait a minute!" I objected.
>
> "Weinberg won't stand in your way," said Rickover. "He knows we need help."
>
> "I don't even know if I want to work here. What would I be doing?"
>
> "Hell, I don't know. That's what you're supposed to help us figure out. Look, if you stay at Oak Ridge, five years from now things will still be pretty much the same. If you come with us, you'll help us build the world's first atomic engine. If that doesn't excite you, you don't belong here."

Life With Rickover

Of course, I signed on. Rickover's team seemed very small, but I gradually found that some people were away taking the Oak Ridge course, and others were overseeing the work at the Westinghouse Bettis lab being built near Pittsburgh. Some were at the AEC's Argonne laboratory near Chicago, since the AEC had just announced that all reactor development would be done at Argonne, and reactor work started at Oak Ridge would be transferred. (That policy decision was never fully carried out, due to the persistence and wily skills of Alvin Weinberg and others at Oak Ridge, but at the time the policy was unequivocal.) Many of us got enticing offers to work at Argonne but decided to stay on at Oak Ridge, where Rickover found me. So the number of people in the Washington headquarters was small, less than a dozen, but the number varied as people moved in and out, and within the next few months it began to grow. Rickover's people didn't do bench work in laboratories. Our job was to see that the right questions were being asked and that the people in the labs were answering those questions by the time the answers were needed. This technical direction and guidance function was a continuing process.

I soon found myself involved in reactor coolant technology, determining the characteristics of potential reactor cooling fluids and the problems associated with their use in a nuclear reactor. The ordinary chemistry of very hot water was not well understood, and its radiochemistry was completely new territory. The other reactor coolants being considered included liquid metallic sodium and gaseous carbon dioxide, and their characteristics were also largely unknown. A new employee, a sharp young Ph.D. named Tom Debolt, was assigned to handle instrumentation and controls for me, but that subject soon became too big for my little corner of the world, and DeBolt was transferred to a growing section dedicated to that subject. Since the radiological aspects of radiation safety kept growing, I soon had a Navy Captain, Oscar Schneider, a physician in the Medical Corps, and later Commander Royce Skow, a graduate physiologist, under my wing. In the rest of the Navy, the assignment of career officers to a civilian would be highly unorthodox, but such considerations never bothered Rickover. We went around and gave presentations to the President's physician, to the Naval War College, to the Pentagon and Bureau of Ships technical people—wherever people wanted to learn out about radiological aspects of nuclear propulsion.

I found out about naval Fitness Reports when I filled out my first one. Luckily I had the sense to show it to Dunford before turning it in. He blanched when he saw it. "Are you trying to get this guy thrown out of the Navy?" he asked. "What do you mean?" I said innocently. "He's actually pretty good. But all these guys are good. I considered checking the box that said he was a 'typically effective officer,' isn't that pretty good?" I quickly learned that any Fitness Report that didn't describe the officer in question as a cross between Dwight D. Eisenhower and Robert E. Lee would set back his career irreparably.

In March 1950, the Argonne laboratory submitted a report proposing three possible design concepts, and we agreed to proceed with the one cooled by pressurized high-temperature water for the first submarine. That decision really focused our actions from then on. The

General Electric people were working on a sodium-cooled design, but it was on the back burner for the moment. We ruled out the gas-cooled design as just too bulky to cram into a submarine hull.

Westinghouse was rapidly building up its naval reactors laboratory at the former Bettis airport near Pittsburgh, working out of the old hangars while the new facilities were under construction. Rickover had insisted that they bring in some of their best people from elsewhere in the company, and the Westinghouse president, Gwilym A. Price, who had been personally convinced by Rickover of the importance of nuclear power for the future, cooperated fully.

Rickover's offices were something special. Edward R. Murrow, the legendary reporter and TV interviewer, began his foreword to Rickover's book *Education and Freedom* (E. P. Dutton, 1959) with these words:

> Washington still has two kinds of office buildings, the roomy modern kind worthy of a great capital and those thrown up as emergency quarters during the war, which by now are dingy and slum-like. In a rundown two-story back-alley building, well behind Constitution Avenue, is housed a staff of men who have done, and are doing, one of the extraordinary jobs of modern times ... These are the offices of Vice Admiral Hyman G. Rickover and his men.

For some reason, the office conditions we worked under didn't strike me as unusual at the time. But they were remarkably grungy—not like any office you would see today, even in some underfunded agency. We didn't lack money, but Rickover believed in austere living conditions. The corridors were long and gloomy, and the dark brown linoleum on the floor was wearing thin. In these temporary buildings it was the government's practice to install patches in the thin places of the linoleum, taking care to match the color and texture as closely as possible. Rickover delighted in this practice. He insisted that the patches be bright red, and he ordered that they be in the shape of boats, or coffins or other provocative forms. Where holes in the floor had to be covered, the patches were not linoleum, but galvanized iron, held down with carpet tacks. This reinforced his image of poor but honest. The final touch was a bullet hole in the floor where a jealous husband had taken a couple of shots at his wife on the floor below (*not* part of Naval Reactors offices!). Rickover took pleasure in showing visitors that historic landmark.

The window air conditioners didn't work properly in the summer, and let in cold drafts all winter. So every spring they went to the shop to be repaired; a process that usually consisted of merely replacing the refrigerant that had leaked out over the winter. We got them back in the fall, just in time for winter again.

The Captain's office was an unforgettable sight. In his outer office were up to four secretaries, all working at a harried pace. Inside, the walls and ceiling and even the door were covered with classic Celotex perforated soundproofing tiles. There was no sign of pomp or rank. No potted palms or even the requisite American flag. The desk, tables and chairs were all well-used civil-service standard issue. Reports, books, articles and papers of all kinds were overflowing from the bookcases and covered every horizontal surface: tables, chairs, and the floor. Mysterious knickknacks were everywhere; ashtrays with various logos, metal and ceramic specimens of every description, rolled-up drawings and blueprints, and photographs of people and equipment, some framed, some not. Two quotations were framed on the walls:

Heaven is blest with perfect rest,
but the blessing of earth is toil

and

OUR DOUBTS ARE TRAITORS
AND MAKE US LOSE THE GOOD WE OFT MIGHT WIN
BY FEARING TO ATTEMPT.

At unannounced intervals, one of the secretaries would walk in briskly and hand the Captain a note. He might say, "I'll call him back," or he might get on the phone and bark a brief message. Sometimes he would delegate, ("Give it to Dunford," or Kyger, or whomever)." Oftentimes he'd order, "Wait outside. I have to take this call." And the secretary would quickly escort the visitor into the outer office, shutting the inner door just as the Captain's voice was reaching a new level of intensity. It was an unnerving experience.

The Role of Paperwork

One of the first things that happen when a project moves from the scientific research stage to the engineering development stage is that the engineers begin to get things down on paper. Scientists generally pass along orally to each other the lessons learned. During the research phase, such oral messages often get informally edited and interpreted, and sometimes this leads to new discoveries and better ways of doing things. But by the time you get to the engineering phase, there are a lot of little details that must be kept straight, and it becomes necessary to write them down.

This applies to every aspect of the work, starting with the raw materials. This was illustrated by an incident that occurred when Admiral Rickover was at the Bethesda Naval Hospital recovering from a heart attack. The senior physician at the hospital was a pioneer in using metal pins for setting bone. Although this didn't apply to Rickover's case, he never passed up a chance to learn. He discussed the subject with the doctor, who told him they were amazed at how much one body's reaction differed from another's. Posthumous examination showed that some pins remained shiny after 20 years, whereas others started to corrode and deteriorate within weeks.

"How do you specify the stainless steel?" asked Rickover. The doctor replied that he specified the dimensions to the degree of precision necessary in each case.

"No, no, the material. How do you specify the grade of material?" asked Rickover. The doctor said that was usually left to the purchasing people. Rickover then had his top design and materials engineer, Harry Mandil, bring in Naval Reactors' stainless steel material specifications: hundreds of pages of fine print, numbers, tables, and references to other specifications, discussing chemical composition, heat treatment, surface condition, cleaning methods and many other factors.

"This is the only way you can be sure of what you're getting," said Rickover. "Until you're sure the material is the same every time, you can't assume it's the body chemistry that's making the difference."

Formalizing Procedures

Procedures, as well as materials and designs, must also be formalized. If you have seen movies of pilots or astronauts going through pre-flight checks-offs, you know how this works. One person has a clipboard with a list of items to be checked, along with the permissible range of numbers for each. He calls off the first item, and the person watching the instrument panels calls out the reading. If the instrument reading is within range, the person with the clipboard checks it off and calls out the next item, about like this:

> "Engine number one, oil pressure?"
>
> "Sixty-five P.S.I."
>
> "Check. Oil temperature?"
>
> "Two hundred degrees."
>
> "Check..."

And so on. The point is that the procedures for normal operation, various emergency situations, maintenance and tests, shutdown, startup and standby conditions and every other foreseeable situation are all written out by design and analysis people, not by the operators. This aspect of Rickover's operation was original and unique. The usual practice is to have *operators* write procedures based on "how it's usually done." But design and operation are two very different skills, and by having designers write the procedures and operators propose any necessary revisions based on hands-on experience, the best of each is incorporated. The operators should have available to them all the background and foresight the designers can write into the operating manuals. After a written procedure is put into use on the plant, something unexpected might happen. The operators can then report the problem back to the designers (like astronauts calling "We have a problem, Houston"). The designers might then have to make some additional calculations or computer simulations before they could write up a revised procedure.

It is important to note that in our program, unlike most industry, the nuclear plant operators were so skilled and so knowledgeable of the technical principles underlying their operations that they were able to suggest significant changes and clarifications in the procedure, and they were strongly encouraged to do so. Many of these suggestions were incorporated into the final procedures. But operators were never permitted to decide independently to modify or ignore a procedure. Besides normal operations, we also needed procedures to cover testing and preventive maintenance, defining in detail how these were to be carried out, how often, what results were acceptable, and what to do if acceptable results were not obtained. This approach is basic to any engineering enterprise and is intended to minimize the chances of one person's mistake leading to real trouble that another person's knowledge could have prevented.

The engineering profession has a tradition of calling on the whole engineering community in a number of ways. Engineers employ the same means as those used by scientists—technical articles, books, seminars, conferences, workshops, in-house information meetings and the like—to keep up with cutting-edge developments and innovations. This is a very active part of their working lives and occupies a great deal of their time. By keeping their contacts open, engineers also learn to profit from the problems and experience of other individuals. There are two different kinds of communities that can be helpful. First are people in the same field, such as the American Nuclear Society, the Electric Power Research Institute, etc. The second community includes people in various fields whose experience with a particular piece of equipment or analytical technique might be applicable. Examples include The American Welding Society, the Institute of Electrical and Electronic Engineers, and the manufacturers of pumps, valves, heat exchangers, instrumentation and controls.

The engineering field also has another more formal mechanism for exchanging information across project lines, the system of codes and standards. There are literally hundreds of codes and standards committees, involving thousands of engineers, each spending hundreds of unpublicized and uncompensated hours a year. This practice got started early in the twentieth century when pressure vessels and steam boilers began blowing up at an increasingly alarming rate. Technical people from the Hartford and other insurance companies started getting together with leaders of the American Society of Mechanical Engineers to discuss how to stop the slaughter. Out of this developed the Boiler & Pressure Vessel Code, various welding codes and standards, and a whole spectrum of metallurgical, mechanical, chemical, electrical, and nuclear standards, now integrated through the American National Standards Institute. Members of these committees are volunteers from universities, research institutes, manufacturers, consultants, and government agencies, whose employers agree to continue their salaries while they carry out this *pro bono* work. Many are retirees, with a lifetime of professional experience to offer. The committees themselves, though, are completely independent of government or of any particular industry. They sometimes draw up recommended legislation, which states or the federal government may choose to copy or use as a basis for further investigations of their own. These standards committees are a unique resource to the engineering profession. This sort of work occupied a great deal of the best Naval Reactor people's time.

The overall engineering philosophy is summed up neatly on a poster I once saw in a tiny, light-plane airport near the Idaho-Montana border:

The Superior Pilot
is one who uses Superior Judgment
to avoid situations
that require Superior Skill.

Graphing Numerical Information

Engineers and scientists have developed a valuable technique for extracting meaning from a mass of numerical data. They find various ways to present the information pictorially in the form of graphs. This is a very effective tool, which I often use. When created insightfully, graphs may reveal trends, cycles, relative magnitudes and deviations that would be hard to discern from tables of numbers. Some of these graphs are quite sophisticated, but the basic idea is simple. We have all become accustomed to seeing bar charts, pie diagrams, and the old-fashioned bumpy line-graph that cartoonists use to depict a rapidly failing business or the chart at the foot of a hospital bed. These seem to convey meaning at a glance. But there's a catch: graphs can also make it easy to mislead. And it is one of my pet peeves as an engineer. It's a problem important enough to pause and examine for a moment.

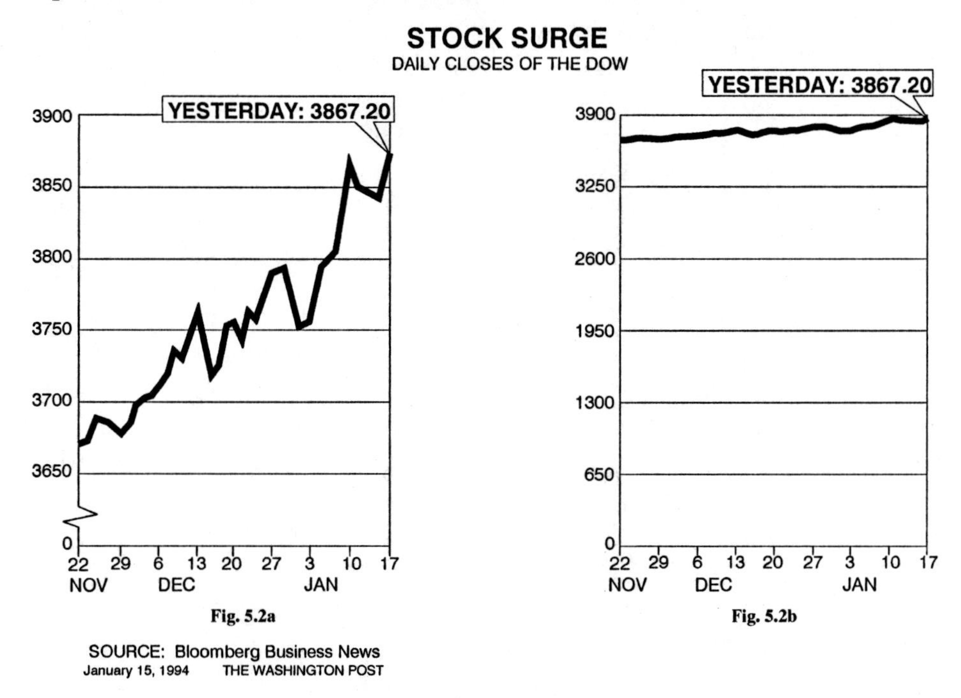

Figure 5.2 How to Make a Slight Gain Look Big

Figure 5.2a from *The Washington Post* shows a "stock surge," and it certainly appears to be a big one. The line depicting stock prices has jumped from 1½ lines above the bottom of the graph to 5½ lines, a 267 percent increase. Wow! Happy days are here again! But there's something wrong. The numbers at the left-hand side of the graph say that the prices increased from about 3,675 (about halfway between 3,650 and 3,700) to 3,867, an increase of only 5 percent. What's going on here?

The numbers on the left side reveal the secret: there is a little jagged break in the scale between 0 and 3,650. All the lines between 0 and 3,650 have been left out, 36 x 2 = 72 of them. That first point on the curve is not 1½ lines above the bottom, but actually 73½ lines above the bottom. The top point is 4 lines higher or 77½. The difference between 77½ and 73½ is about 5 percent, just as it ought to be.

This graph violates one of the most basic rules for graphical portrayal of data. By breaking the vertical scale, it distorts the amount of change or variation. It makes a 5 percent change look like a 267 percent change. And that's just not honest. With this trick, any graph can be made to look as steep as you wish. You could show a change from 1,001 to 1,002 as a leap from the bottom of the paper to the top, just by omitting the first thousand lines. I have tried to get editors to stop this dishonest practice, but they protest that doing it properly would make many changes appear so small as to be trivial. But if that's what the numbers say, the graph should not tell a different story. Editors know that they should not make such distortions with *words*. I maintain they should not create such distortions with numbers either. Don't be taken in—just look for the zero. If there are not equally spaced intervals all the way from zero to the top of the chart, then you know you are seeing a distorted picture of the data. There are some situations for which this is appropriate (such as data ranging over several factors of ten), but then the presenter should make very clear what has been done.

Review and Evaluation by Outsiders

In addition to drawing on help to develop designs and procedures, engineers working on public projects such as nuclear power plants undergo technical scrutiny from people outside the organization. In the civilian nuclear power field, the Institute of Nuclear Plant Operations (INPO) serves as an industry watchdog made up mostly of experienced operators and technical personnel from other nuclear facilities. The Nuclear Regulatory Commission (NRC) is the official federal oversight agency. In addition, there is the statutory Advisory Committee on Reactor Safeguards (ACRS). We'll talk more about these organizations later.

Rickover the Enigma

Admiral Rickover gave every appearance of being the ultimate technocrat. He took obvious personal delight in digging into every technical detail and worrying it to death. He repeatedly expressed public concern at the low state of scientific literacy in the country and the failure of public education to correct this deficiency. This is the sort of behavior we associate with persons who find no pleasure in social activities. And Rickover fitted that image too. He had little to do with parties and gossip and other people-centered things. Although he loved to find someone who knew things he didn't know and eagerly absorbed that new knowledge, he did not seem to enjoy the company of people for company's sake. He got no joy from small talk. All that fitted the stereotype.

But it is also true that Rickover always considered individuals to be the overwhelmingly critical factor in everything he did. He knew if he got the right people working for him, he had the best chance of solving whatever problem came along. Even before there was a nuclear power program, he persuaded Oak Ridge and MIT to set up special nuclear engineering education programs to ensure the availability of trained professionals and technicians for any circumstance. Later, he insisted that his laboratories and shipyards take the same attitude, and he gave them the resources to do so. He took extraordinary pains to develop, train, sustain and retain his people at all levels, in contrast to the attitude of so many managers who view their people as easily replaceable. In any new situation involving other organizations, he always looked carefully at the people he was about to deal with, whether they were a new contractor, a new government agency, or a new congressman. He wanted to know who held the important cards, because if he could get that person on his side, or alternatively, if that person turned against him, this could become more important than any other he knew everything else might become secondary.

Despite the trust and responsibility Rickover bestowed on his people, he personally involved himself in the important details and most of the unimportant ones as well. He made the critical design decisions by chairing a technical discussion with all involved parties from headquarters, the reactor design laboratory, the shipyard and vendors. And he never missed an initial sea trial except when he was in the hospital.

By the fall of 1950, Rickover had what he needed to begin. People that he had carefully selected were beginning to report in from various intensive training programs. On August 8, 1950, President Truman had authorized a nuclear submarine in the 1952 shipbuilding program, with a January 1955 initial sea trials date. (Rickover had set this date, but it horrified the ship designers in the Navy's Bureau of Ships, who considered it much too optimistic.) Rickover had line authority in the AEC's Reactor Development Division to develop the propulsion plant, and Westinghouse was building the Bettis Atomic Power Laboratory to carry out that development program. Rickover also had line authority in the Navy's Bureau of Ships to design and build a propulsion plant in the ship now authorized, and the Electric Boat Division of the General Dynamics Corporation had been contracted to build the full-scale prototype propulsion plant in Idaho. The contract for the ship itself was still a year away. Although the essence of the design was just coming into focus, Rickover ordered Electric Boat to begin work on the full-scale prototype plant in Idaho. The countdown had begun.

6. Radiation, People, and the Good Earth

Radiation: The New Wrinkle

It was clear from the start that there were many aspects of building the first atomic power plant where we could be guided by experience with other power systems. But the presence of radiation made a big difference. No other power plants had dealt with this problem before. Not only was there no past experience with designing and building radiation shielding systems for ships or power plants, there were no precedents to tell us what acceptable level of radiation we should specify for the crew to receive.

Rickover was a very conservative engineer; if he could have specified *get it down to zero,* he would have. But he could not, for two very good reasons. First, because shielding reduces radiation *exponentially*; that is, two inches of lead will reduce gamma radiation ten-fold, another two inches another ten-fold, but no amount of shielding will reduce it to zero. So if you had a thousand gamma rays coming out of the reactor every second, you could reduce the number to one hundred, or to ten, or even to one gamma every second. You could get it so low that your instruments couldn't detect it. But you couldn't get it to zero.

This idea of exponentials is key to understanding nuclear power that we need to digress for a moment to discuss it. (The second reason you can't get to zero radiation is because the whole world is naturally radioactive. I'll talk about that in the next chapter.)

The story of radiation, like most of science and technology, is primarily a numbers game. Like salaries or the scores of ballgames, you don't get to the real meat of the subject until you start quoting numbers. Admiral Rickover put it thus: "The Devil is in the details—but so is Salvation." In the nuclear game, *details* generally means numbers. So it's important to understand just what kinds of statements with numbers are meaningful, and what kinds of statements are nonsense or worse.

Powers of Ten

Because scientists and technologists often have to deal with very large or very small numbers, they have come up with a simple way of expressing them. They express all such numbers in terms of *exponents*, or *powers of ten*. Thus, 100 is 10 x 10, or *10 to the second power*, written 10^2. One thousand is 10 x 10 x 10, or 10^3 · Note that in this example, the power of ten is equal to the number of zeros.

Thus, scientists write one million as 1×10^6, and they write two and a half million as 2.5×10^6. Although 2,500,000 has only five zeros, and 2,533,700 has only two zeros, they are not different powers of ten. They can be written as 2.5×10^6 and 2.5337×10^6, respectively. This way of writing numbers is not new to you. News reporters describe a $3.87 billion expenditure or a $1.4 trillion national budget. A scientist, however, would write these numbers as 3.87×10^9 and 1.4×10^{12}.

After you get used to it, this is really a lot simpler than working with *quadrillions* and *quintillions* and *millions of billions*. The numeral 10^{12} means the same thing all over the world, even though Americans call it trillions and Europeans call it billions.

The same sort of system can be used for very small numbers; we just use *negative* powers of ten. For example, one-tenth is written as 10^{-1}, one-one thousandth is 10^{-3}, and so on. Using this notation, we can describe the smallest distance within the nucleus of the atom (which is 10^{-35} inches), to the largest number that we can assign any physical meaning, the total number of atomic particles in all the atoms in the universe (thought to be about 10^{75}). This range of numbers is almost unimaginably greater than the range of numbers most of us deal with in our daily lives. The smallest speck we can see might be 10^{-4} inches, while the height of the highest mountain we can view would be perhaps 3.5×10^5 inches. Figure 6.1. puts these numbers in perspective.

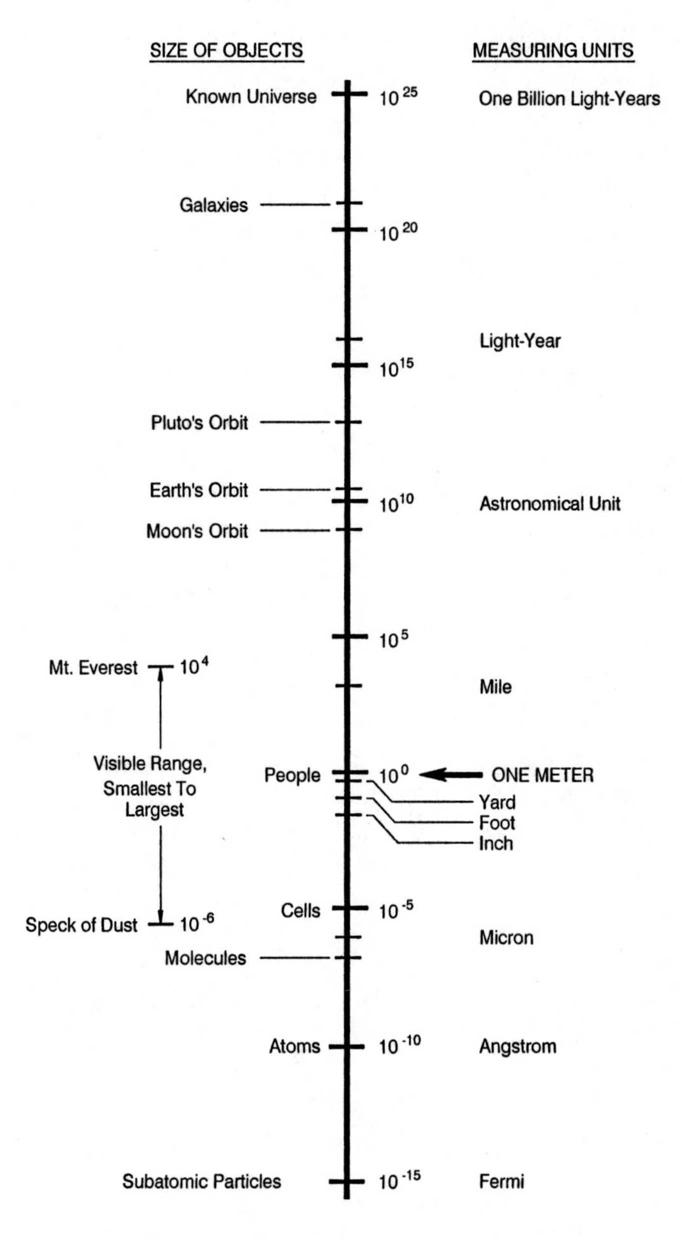

Figure 6.1 Powers of ten chart (*Hoke*)

Precision and Accuracy

Using numbers allows communication to be a lot more precise, with less chance for misunderstanding. If I predict that it is going to rain some time next month and it does, my prediction was *accurate*, though not very *precise*. If I predict that it will rain Wednesday afternoon and it rains Monday, my prediction is more precise but not accurate. Whether my accuracy results from sophisticated knowledge or sheer luck is not relevant here. The question is whether I hit the target. *Accuracy* is hitting the target. *Precision* defines the size of the target. By working within numerical i.e., measurable, limits, we know a great deal more than if we used only vague, qualitative words such as *very soon*, or *after a while*.

Of course, there are times when it's acceptable to be imprecise with numbers. We say, "He's a *millionaire*," or a *multi-millionaire*, or (since the 1980s) a *billionaire*. Or we say her salary is *six figures*. These are perfectly proper ways to state the case, but they are still numbers. Six figures means more than $99,000 (five figures) and less than $1,000,000. Even low-precision numbers are more informative than words like large, huge or humongous.

Use of numbers enables us to tell exactly what degree of precision our information entails. If I say a bucket holds 7 gallons of water, that means I know that it holds more than 6½ and less than 7½ gallons. If I say it holds 7.0 gallons, that implies that I know it holds between 6.95 and 7.05 gallons. (If it's less than 6.95 then we'd call it 6.9, not 7.0.) So I can convey some precise information with a very simple notation system.

Suppose I lend my bucket to a foreigner. Nearly all the industrialized countries of the world except the United States use the metric system, so this person might want to know how many liters my bucket holds. I look up the conversion and find that a liter is 1.0567 quarts, so my 7-gallon bucket holds (as read from my calculator):

$$7 \times 4 \text{ (quarts)} \div 1.0567 = 26.4975868269 \text{ liters.}$$

But if I tell people that this is how many liters it holds, I am misleading them. I would be implying that I know the capacity of the bucket to one part in 10^{12}, for that is the precision of this long number.

If in fact I know only that the capacity is 7 gallons, not 7.0, then I know the answer only within a half-gallon, or about 2 liters. I should tell the borrower that I know the bucket holds 26 plus or minus 2 liters, or I could say it holds between 24 and 28 liters.

So by using numbers properly, an engineer or a scientist can convey quite a bit of information, in a simple and unequivocal way. There's nothing mysterious about it. We do it for very practical reasons.

Radiation Standards

Rickover and his people were not the first to deal with radiation. In the closing years of the nineteenth century, Pierre and Marie Curie began separating uranium, radium and other radioactive materials from a natural ore called *pitchblende* that gave off, as they concentrated it, "a warm glow in the evening," bright enough to read by. About the same time, Wilhelm Konrad roentgen discovered x rays. Shortly after his announcement, scientists all over the world found they could also rig up equipment to produce x rays. In a matter of months, x ray equipment was available for $15 and being used, among other purposes, in beauty parlors to remove unwanted hair.

The public reaction when x rays and radioactivity first appeared was one of optimistic excitement. John Lenihan, Professor of Clinical Physics, notes that various "cures" using radioactivity were advertised: girdles, contraceptive jelly, hair tonics, chocolate bars, tooth paste and hearing aids, all loaded with radioactivity, were said to be boons to health. The first warnings about x rays were concerned with morality rather than health hazards; the concern focused on the possibility that mischievous people would carry on improper anatomical inspections. *Photography* magazine wrote:

> I'm full of daze,
> Shock and amaze;
> For nowadays
> I hear they'll gaze
> Thru' cloak and gown—and even stays,
> These naughty, naughty roentgen rays.

A British newspaper advertised x ray spectacles "that give you the amazing illusion to see right through everything you look at ... the most amazing things when looking at girls and friends!" As an antidote, papers advertised x ray proof underwear. The New Jersey legislature considered a bill to prohibit the use of x rays in opera glasses.

At first, little thought was given to protecting people from the radiation. However, after Edison and others reported skin burns from the new rays, efforts were begun to measure the radiation intensity and to devise protective standards and procedures.

The first thing that was needed was a unit of measurement. You can't measure radiation in inches or in quarts, but you can measure how much energy is emitted from a radioactive source, and how much energy is deposited in tissue by a given field of radiation. So two units were needed: one for *radioactivity*, and one for *radiation*. The first unit of radioactivity was called the *curie*. It was based on the radioactivity of one gram of radium. Any source of radiation with that amount of atomic activity was said to be one curie. How much radiation does that give you? It depends on how close you stand to the source. The amount of radiation measured one meter from one gram of radium is about one *roentgen* per hour. The curie and the roentgen were defined more elegantly, but this is a good way to remember it. One curie of radium gives one roentgen per hour at one meter. If you back off far enough, you won't be able to detect any of its radiation at all.

Different kinds of radiation cause different amounts of damage. Low-energy x rays can burn the skin but cause little internal damage. High-energy gamma rays deposit their energy almost uniformly through the body. So the idea of *relative biological effectiveness* was developed to account for the different kinds of biological damage that the same amount of energy can inflict when it comes from different kinds of radiation. Neutrons may be up to ten times as damaging as gamma rays, and alpha particles up to twenty times. For radiation protection work we have a unit of measurement called the *rem*, which was derived from *roentgen equivalent, mammal*. One unit of energy from neutrons will give about ten times the dose in rems as one unit of gammas. Of course, choosing a single number to compare the relative biological effectiveness of different kinds of radiation is quite arbitrary. How bad must a burned hand be to be equivalent to a sore throat, diarrhea, or a migraine headache? But the rem is the best we have, and it serves the purpose. Ironically, the word *dose* is used to describe the amount of radiation absorbed, because the first uses of radiation were for healing.

Metric Units

Starting in 1975, in an effort to get a consistent basis for all units of measurement based on metric units of weight, energy, volume, etc., metric units were officially adopted for measuring radiation and radioactivity. Americans have been slow, however, to convert from units with which they have grown comfortable after a century of use such as inches, pints, and pounds. U.S. regulations are all still written in American units. I will continue to use the American units in this book, although you may see other authors use units named after L. H. Gray, Rolf M. Sievert and Henri Becquerel, early researchers in the then-new field of radiobiology.

All these units, and all the discussion on radiation in this book, apply only to what we call *ionizing radiation.* That's what is emitted from radioactive materials and from the fission process. Other kinds of radiation, such as sound and ultrasonics, radio, infrared, sunshine, ultraviolet, microwave ovens and the radiation from electric power lines and cellular telephones are all called *non-ionizing radiation* and affect living tissue very differently. That's a whole other book.

Genetic Damage

There is only one thing more fearsome than the thought of our bodies disintegrating at the cellular level, and that is the thought that future generations might be irreversibly deformed through damage to the DNA. Could radiation alter the very genetic structure of the cells we pass on to future generations?

Genetic damage is a subject that many people associate uniquely with radiation. But the fact is that nuclear radiation is not particularly effective in causing genetic damage. When scientists want to cause genetic damage to fruit flies or mice or other organisms in the laboratory, they usually choose chemicals, ultraviolet light, or other agents more effective than nuclear radiation. In addition to our theoretical and laboratory knowledge of the effects of radiation, we have some direct evidence from people irradiated at Hiroshima and Nagasaki and their descendants. They were subjected to much higher radiation levels than those associated with nuclear power. Although an international team of physicians has been studying these survivors and their children for more than fifty years, and is now looking at some grandchildren of those survivors. They have found that the radiation has not affected subsequent generations, i.e. the number of birth defects in the descendents of irradiated persons does not exceed the number found in unirradiated control groups. (Note that this does not include the damage done to some fetuses directly irradiated in the womb by the explosion. That was not genetic damage; it affected only the organism irradiated and does not transmit to subsequent generations.)

In addition to information on atomic bomb survivors, we have data from regions of the earth where the natural radiation level is much higher than average—for example, Colorado, Brazil and several locations in India and Iran. We find that people living there do not have any more occurrences of genetic defects or unusual health problems than do people living elsewhere. In fact, the Colorado Plateau states have a natural radiation level *several times* the average of the American gulf coast states, but their cancer rate is significantly less. This is because radiation is not a major cause of cancer.

"Fouling our Nest"

The human gene pool is not being degraded by radiation, but what about the earth? People ask, *The amount of radioactivity you add to the earth's burden each year may be small, but aren't we then continually and irreversibly increasing the overall radiation background?* The answer is no, for two good reasons.

First, the total amount of radioactivity generated by human activities is small compared with the earth's natural radioactivity. The natural decay of earth's natural radioactivity more than compensates for it. More important is the fact that we are taking the fissionable isotope of uranium, with its nearly billion-year half-life, and replacing it with fission products most of whose half-lives range from seconds to a few thousand years. So in the long term, we are actually *decreasing* the earth's overall radioactivity, although not enough to make any difference in human life spans.

Fear of Radiation

Because nuclear energy first appeared on the public stage in the fiery demolitions of Hiroshima and Nagasaki, we expected that the peaceful application of that energy would be perceived publicly with some apprehension. Through the 1950s and '60s, this apprehension was scarcely apparent. We sounded out public opinion from time to time around our facilities, but found little concern. However, as the various protest movements of the '60s gathered strength, nuclear power was swept into the target area alongside nuclear weapons, and fear of nuclear weapons fallout extended to reactors as well as bombs. It was easy for us to blame the media. They seldom used the word *radiation* without preceding it by *deadly* or *lethal*, despite the wide use of radiation to treat cancer and save lives. Gravity, water and even air kill far more people (through falls, drowning and hurricanes) than radiation. But we seldom read about "deadly gravity" or "lethal water." This routine demonization of radiation seemed inflammatory to us, but most people seemed to accept it unquestioningly.

Anti-Nuclear Hysteria

I remember a reporter expressing great concern over the risk of cancer to uranium miners. When I pointed out to her that the miners were ten times more likely to be crushed to death than to die of cancer, and that overall they were at lower risk than coal miners who often die of black lung disease, she could not understand how one could talk of "ordinary dangers" in the same breath as accidents involving radiation. "*Nothing* is as bad as cancer or radiation sickness," she insisted, implying that the two were the same. She didn't want to listen to anyone knowledgeable about radiation, "because they're biased."

Of course, the only alternative to being biased about any subject is not knowing anything about it. One's knowledge may come from scientific data and hands-on experience, or it may be based on rumor and fear mongering. But there is no way to escape bias except through total ignorance—not a good alternative. We just have to make sure that our biases do not cloud our selection and interpretation of the facts. And there is an abundance of facts about radiation.

Dealing With the "Anti-Nukes"

Looking back, it's clear we never really engaged the nuclear critics in rational debate. It seems we were always talking past each other. I'll admit I get a little testy when I read, even today that, "the scientific and technical community that did not foresee the present situation now finds itself trying to figure out how to store nuclear waste." We have been working on these problems—and solving them—for decades. Starting with Einstein's letter to President Roosevelt in 1939, before many of the protesters were born, the dangers that could result from an improperly controlled atomic energy program were the primary focus of most of us in the atomic energy field. Those of us who would control that energy for peaceful purposes were always painfully aware of the need to ensure that we could do so safely. Perhaps, in retrospect, this is why our answers to the protesters may have appeared dismissive. We were offended that they were presenting themselves as the first to perceive safety problems. In fact, another whole generation before the discovery of fission—*two* generations before most of the anti-nuclear protests began—medical scientists were developing radiation protection standards and procedures for safely applying x rays and radium for medical and industrial purposes.

Some people act as if fear of radiation was a mark of a good environmentalist, but in fact a number of prominent environmentalists have spoken out against this idea. Stewart Brand, former editor of *The Whole Earth Catalog* and *CoEvolution Quarterly*, and Ansel Adams, the noted nature photographer, have been particularly explicit in that regard.

James Lovelock, foremost advocate of the Gaia Hypothesis that the earth is in effect a living organism, wrote simply: "I have never regarded nuclear radiation or nuclear power as anything other than a normal and inevitable part of the environment." He went on to note that the radiological damage to human tissue is caused by the same chemical decomposition products that are produced by normal breathing. "...In other words, so far as our cells are concerned, damage by nuclear radiation and damage by breathing oxygen are almost indistinguishable." Putting in the numbers, he concludes, "Breathing is fifty times more dangerous (in destroying cells) than the sum total of radiation we normally receive from all sources."

Lack of Public Understanding

Being ignorant or misinformed about radiation is not just quaint; it can actually be life-threatening. It is estimated that about 100,000 needless additional abortions were performed in Europe after the Chernobyl accident, presumably because the parents feared that radiation might have deformed their child—a fear totally without scientific basis.

Each of us has to make decisions from time to time that require a basic understanding of radiation. We have to accept the use of x rays or other diagnostic or therapeutic uses of radiation for instance, or else we must choose alternatives that may be dangerous to our health. The use of radiation and radioactivity for diagnostic and therapeutic purposes is saving thousands of lives each year, yet irrational fear of radiation leads many people to refuse such healing treatments. Here are a few examples, starting with some cited in 1995 by Professor Bernard Cohen, a radiation expert at the University of Pittsburgh:

- In one major hospital, about 20 percent of the patients refuse the use of radioactive iodine treatment for hyperthyroidism, opting instead for a less effective drug treatment that often leads to relapse
- In another hospital, whenever a portable x-ray unit is brought into the intensive care unit, the nurses leave the area, abandoning the infants under their care, although the dose they

would have received if they had stayed would have been completely harmless.

- A large medical center planned a project involving radiation for 100 patients a year. The ten nurses working in the unit threatened to strike, although again, the radiation they would have received from the project would have been insignificant. The project was abandoned.
- Despite evidence that early detection by mammograms is the best defense against lethal breast cancer, the leading cause of cancer deaths among women, more and more women are refusing to have mammograms, solely because of a groundless fear of the radiation involved.
- Ralph Nader and others campaigned against smoke detectors that use a small radiation source, on the basis that "any amount of radiation is harmful," a statement without scientific merit. There have been crusades against irradiated food, which could have prevented deaths from salmonella poisoning.

Some people fear their microwave ovens because of what they have read about ionizing radiation, although the microwave radiation is of an entirely different kind, with unrelated biological effects. In two extreme examples, two nurses at a Raleigh, North Carolina, medical center refused to treat the wife of a vice-president of the local utility on the grounds that the utility was building a nuclear power plant. In the other, a well-qualified student was denied admission to graduate school because in the eyes of most members of the student admission committee he had a repugnant characteristic: he was an engineer at a nuclear power plant. These, of course, are political actions, but they are based on the unique fear that radiation evokes in many people.

Despite all these fears, there is no evidence that anyone in the U.S. public has ever died from exposure to nuclear radiation from a nuclear power plant. Some persons have hypothesized that over a large population some number of deaths may have been due to radiation, but these are unwarranted speculations.

"Has Radiation Protection Become a Health Hazard?"

Gunnar Walinder, prominent Swedish radiobiologist, wrote a book with that provocative title. As a former Chairman of the Swedish Radiobiology Society and senior member of a number of international professional and policy-setting committees on radiation standards, he is well qualified to consider the question. He concludes that the presumption that any radiation level, no matter how small, poses a biological hazard requiring protective action is not supported by the scientific data and leads to actions that "have been judged by competent observers to have caused more harm than the radiation itself." This is the Linear No-Threshold (LNT) model of radiation response.

In protesting the use of the LNT model, Dr. Walinder is joined by many other well-known authorities including Zbigniew Jaworowski, former Chairman of the United Nations Committee on the Effects of Atomic Radiation; Lauriston S. Taylor, former President of the National Council for Radiation Protection and Measurements; Myron Pollycove, recipient of the Wilhelm Roentgen Radiology Centennial Award; Ralph Lapp, radiation physics pioneer; Thomas D. Luckey and Bernard Cohen, professors of radiobiology, and many others who are urging that the no-threshold concept be replaced by a more defensible model. The French Academy of Sciences published a report on this subject concluding, "there is no scientific basis" for using the linear no-threshold hypothesis as a basis for protective action.

James B. Muckerheide, the Massachusetts State Nuclear Engineer and co-director of the Center for Nuclear Technology and Society at Worcester Polytechnic Institute, has been rounding up this scientific research during the last few years and arranging for seminars to discuss it. He found that assuming "no amount of radiation is harmless" has led to actions that are not only contrary to the scientific data and theories but often also work against the public good. He charges that spending billions of dollars to clean up remote sites, especially those with radioactivity already below the natural background level of many healthful locations such as Colorado, not only wastes money, but also works against the public health. By diverting money needed to clean up polluted air and water in populous areas, these funding decisions indirectly imperil the health of the public. And the remedial work—digging up and removing thousands of truckloads of "contaminated" soil, which poses no hazard, and moving it across the country to somebody else's backyard—can be seriously damaging to the environment.

Applying the Principles

In the early 1950s, realizing that we couldn't get away from the earth's natural radiation, Rickover and his people knew that they could not hope to design the nuclear submarine to achieve a zero radiation level. The questions then became: What was a reasonable design basis? and, How much radiation would be acceptable? Some general guidelines had been developed during the previous decades for the amount of radiation considered acceptable for persons working with x rays or with radium, and these guidelines were applied to workers on the atomic bomb project. At that time, the standard for an acceptable (safe) radiation dose was under heavy discussion. It had been one-tenth of a rem per day, then 15 rem per year, then 12. There was talk of setting it at 5 rem per year, starting at age 18, and that seemed where it might stand. In addition, following common toxicological practice, it was decided that children should not be exposed for long periods to more than one-tenth of the amount of radiation considered safe for workers. Since it was not feasible to protect children specifically, it was decided to apply this lower dose limit to the public at large. This series of lowering numbers was not based on scientific evidence or even concern that the previous numbers were not sufficiently protective. It was just a feeling that "we could do it," and we were determined to leave no basis for claiming that we hadn't been "safe enough."

Since the first nuclear power plant was to go into a submarine, many radiation experts advised us that the civilian worker limits were unduly conservative for that application. Captain Rickover sensed this was a fundamentally important issue, and he personally went to great lengths to get the best possible advice. He talked with the legendary geneticist Hermann Muller, and at Rickover's behest, I talked with health physicist K. Z. Morgan at Oak Ridge. Dr. Morgan and other experts argued that military personnel knowingly risk their lives and, therefore, we should not make the submarine slower and more vulnerable by adding heavier shielding to bring the radiation down to civilian standards when a higher dose of radiation would still be safe. This represented many tons of shielding for a ship where the weight of every item aboard is accounted for to a fraction of a pound. The advice we were getting came from civilian doctors and scientists, people with no particular bias toward the military, and the radiation level they considered safe was quite a bit higher than the civilian standard. (K. Z. Morgan later in life became an anti-nuclear crusader, testifying in countless court cases that even the civilian dose standards were hazardous.)

Rickover did not follow the advice to design for high radiation levels. "If nuclear power ever attains a significant place in the Navy, tens of thousands of sailors will ultimately serve on these ships. Suppose some day, some sailor's wife has an abnormal baby, or whatever," he said. "Such things happen to people anywhere. At that time, I want no basis for anyone concluding that radiation could have caused it. The civilian standards represent the international consensus, set by medical people, of what is a proper and safe level of radiation. We'll build enough ship to float the necessary shielding." Repeatedly situations arose where a reasonable course seemed to call for a liberal interpretation of the civilian standard, but Rickover always insisted that the design meet the most conservative interpretation that might be suggested during the future life of the ship. His sort of reasoning has characterized nuclear power from the beginning.

As it worked out, because nuclear submarines operate submerged most of the time, the seawater and the steel hull shield the crew from most of the cosmic rays, so the radiation the sailors receive, working and living within a few feet of an operating nuclear power plant is, therefore, actually less than the natural radiation they would get if they stayed home. (The members of the crew who have to enter the shielded compartment housing the reactor system for maintenance or repairs during shutdown do receive more than that. Still, the radiation they receive compares favorably with that seen by operators of civilian plants.)

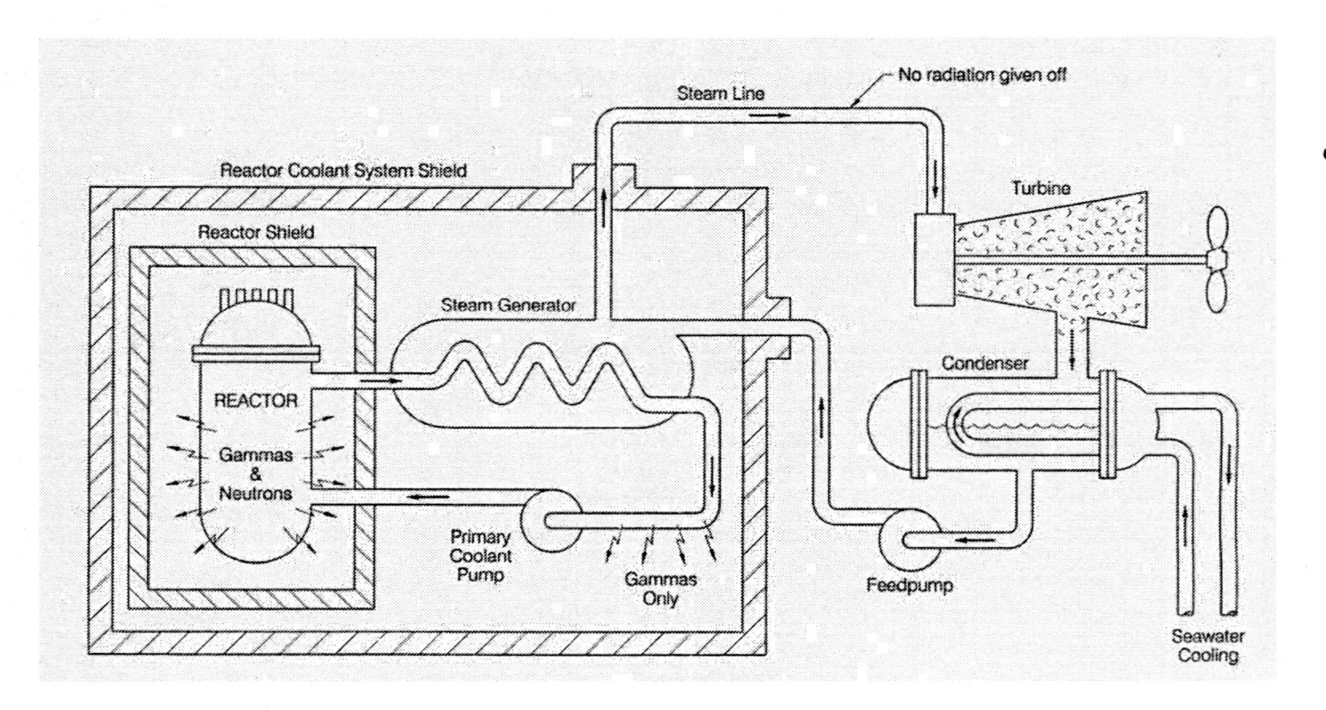

Figure 6.2 Simplified diagram of the submarine shielding problem (*Hoke*)

Designing the Radiation Shield

For the submarine, we started out with a number on which to base the design of the radiation shielding. We used 15 rem per year, but we had enough margin in the design to comply with the 5 rem per year that we felt might be required before the ship was retired.

However, it takes a lot of work to get from determining an exposure limit to a finished design. Let's look at some of the steps involved. We had to shield against two separate sources of radiation. First was the reactor core itself, the uranium-containing fuel elements whose fissions produced the energy that drive the ship. The reactor emitted both neutrons and gamma rays, which are like x rays, only of higher energy. Second, the water that flowed through the reactor to cool it became activated and gave off gamma rays but no neutrons. This water was heated by the reactor and then passed its heat on through a heat exchanger (the steam generator) to a second loop of water, as shown in Figure 6.3a.

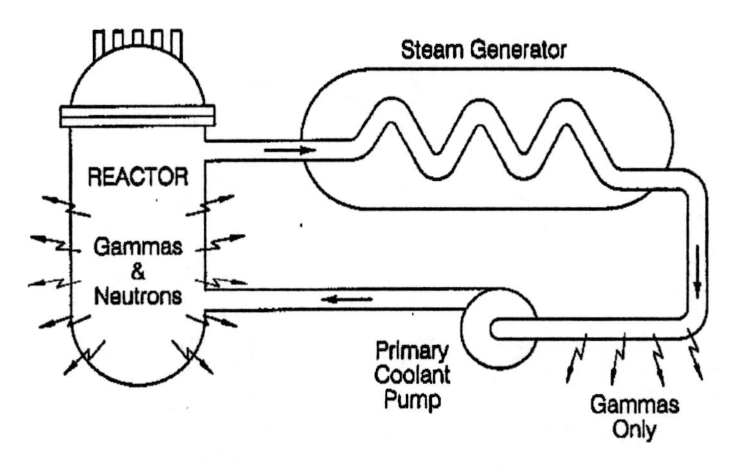

Figure 6.3a The Reactor Cooling System

This secondary water was in a separate piping system and did not mix with the reactor cooling water and, therefore, was not radioactive. Being at lower pressure, the secondary water boiled and became steam, which drove the turbine that turned the ship's propeller. The turbine condenser was cooled by seawater in a third separate loop.

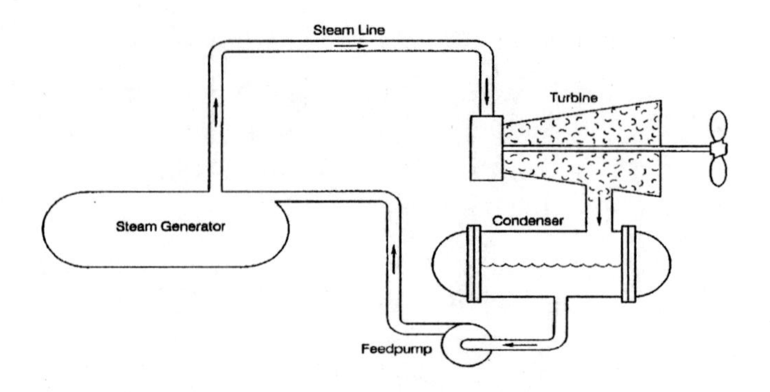

Figure 6.3b The Steam System

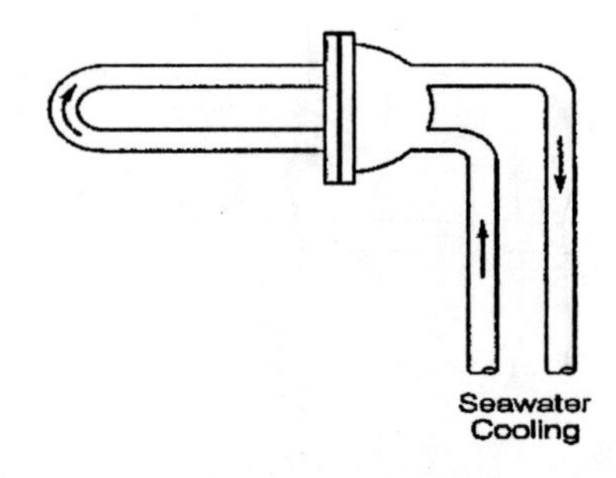

Figure 6.3c The Seawater System

Each of the three loops posed a different shielding problem. For the *reactor*, we had to provide shielding against both neutrons and gammas. For the *reactor cooling water*, we needed only gamma shielding. And for the *steam system*, carrying heat from the heat exchanger to the turbine, and for the *seawater system*, we needed no shielding at all. These three separate loops of water fit together to make the overall plant, shown schematically in Figure 6.2

Of the various kinds of radiation we have to deal with in shielding, *only neutrons can make other materials radioactive*. This is what happens to the water passing through the reactor in the primary loop. But the radiation emerging from this water outside the reactor is essentially all gammas, and gammas do not make other materials radioactive. (That is why irradiated food is not radioactive; it is irradiated by gamma rays, x rays or electrons, not neutrons.) So, although the primary loop is radioactive, it does not make the secondary loop radioactive.

Knowing this, we were ready to begin designing the plant shield. For starters, we pictured the power plant running at full power, and we calculated how much gamma radiation the water in the primary loop would give off. Then we designed a shield around all that piping and equipment to bring that radiation level down to the acceptable level. (We'll get into what *acceptable* means in just a moment.) Still, that wouldn't take care of the radiation from the reactor core, for two reasons. First, because there was a lot more radiation to deal with from the reactor than there was from the cooling water; second, because the cooling water gave off only gamma radiation and the reactor gave off neutrons as well as gammas. So we had to put a neutron shield and additional gamma shielding around the reactor core.

The Shutdown Case

We've talked only about shielding the reactor while it's running at full power. What happens when it's shut down? After a reactor is shut down, the fission products keep giving off radiation, even though the fission process itself has been stopped. That's why there is radioactivity associated with nuclear waste materials. The radiation *from the reactor core* drops immediately to a few percent of its full-power level and keeps decreasing from there, but more and more slowly. On the other hand, the irradiated *water in the piping* decreases a thousand-fold in the few minutes it takes to open up the shielded compartment for maintenance or inspection.

If it weren't for the anticipated small amount of impurities and corrosion products in the water, the radiation level would drop almost to zero. But even with impurities, the radiation level after shutdown quickly drops to a point where people can walk right up to it and work on the equipment. So another shielding requirement became: Let's put enough shielding around the reactor core so that it doesn't add significantly to the radiation already present from the water loop after shutdown.

The big uncertainty here was the amount of corrosion products and other materials that might accumulate in the water. We had little experience to tell us what levels of corrosion products or other impurities might build up and how much radiation they would produce. There wasn't much theory to build on. Only a few parts per million of the wrong impurities could create a major problem.

Getting from Dose to Dose Rate

Getting back to the idea of an *acceptable radiation level* outside of the shield, we figured that the crew should receive no more radiation than a civilian worker, which was 15 rem in any one year. But 15 rem is a radiation *dose*, and we needed to know how many rem *per hour*, or the radiation *dose rate*, to account for in the shielding design. The total dose depends on how many hours each year we expect a person to be near the surface of the shield. And that gets into the specific design features of the ship.

The *Nautilus*, like most of the nuclear-powered submarines that came later, had its engine room immediately aft of the reactor plant, and sonar, radio, radar and other operating stations immediately forward. Some of these locations had manned stations near, but not directly against, the reactor compartment shielding. In fact, we made a special effort to place equipment or storage space against the shield rather than people. A few locations were labeled "3 HOURS PER DAY STATION." A few more were 12 hours per day, but most of the ship's stations had no time restrictions. So we started by assuming that the radiation at the various locations should be no more than 15 rem, divided by 3 or 12 or 24 hours per day. Dividing that rate by 365 days per year comes to a few thousandths of a rem per hour. Rather than deal with such tiny numbers, we usually used *millirem*, or mrem, which are one-thousandths of a rem.

Let's summarize where we've been:

1. *We live in a world of natural radiation that has been here since the earth was formed and has nothing to do with human activity. On the average, this natural radiation field bombards each of us with over one billion radiation particles per day.*
2. *Most radiation goes right through our bodies without harm, but some of the radiation kills cells and some damages cells, which might lead to cancer. However, the number of cells killed or damaged by radiation is far less than one percent of the number of cells killed or damaged in the body's normal process of growth and regeneration.*
3. *Radiation, at the levels encountered in nuclear power plants, does not cause an increase in the normal genetic damage to humans.*
4. *Despite the military mission of their crews, nuclear submarines were designed to give persons onboard no more radiation than was allowed for civilian workers.*
5. *The radiation level permitted for civilian workers was 15 rem per year. For the submarines, we assumed that personnel might have duty stations near the surface of the shield, and those locations were, therefore, designed to give no more than that amount, based on 3, 12 or 24 hours per day, as posted, all at full power. Living quarters were initially designed for 120 hours per day, for extra margin. We included sufficient safety margin to cover possible reduction of the permissible dose during the life of the ship.*
6. *The system that pumped cooling water through the reactor got radioactive while operating and had to be shielded. Its radioactivity decreased a thousand-fold a few minutes after shutdown, so the shielded compartment could be entered for maintenance. The reactor itself was shielded separately, providing for both the full power and the shutdown situations.*

From Concept to Design

It would not make sense to assume for the shield design that a sailor would sit at the very surface of the shield, with the reactor plant running continuously at full power, 40 hours every week, for every week of the year. We designed for a more realistic, but still conservative, situation. First, no watch-station was right against the shield wall. As I mentioned earlier, we always provided some space, and we put some equipment or ship's structure in the space, which cut down the radiation level considerably. On the *Nautilus*, we actually stored food there. Second, there was shore leave and duty elsewhere on the ship, so no one person ever spent 52 weeks a year at any one station. And most important, no ship, even at its busiest, ever operated at full power all the time.

All these factors tended to lower the radiation actually received by the operating crew at least a hundred-fold. This in turn was offset for the engineering crew by the occasional maintenance duty on the shutdown reactor plant. The bottom line is that the crew, like the operators of every nuclear power plant built since then, routinely received only a negligible dose of radiation from their job. (Commercial power plants were not limited by a submarine hull, so they could afford to be lavish with concrete and with distance, and could thus achieve conservative results even with extended full-power operation.)

Design Details

While designing the shield, the biggest arguments revolved around defining radiation specifications for parts of the shield that were not near manned stations. An obvious example is the area directly under the reactor. Nobody could get there, so what if we didn't put *any* shielding there? Just posing the question brought out some answers. Directly below the reactor is the ship's hull, and it does sometimes have to be inspected from the outside by divers and in drydock.

Of greater consequence, while in port, radiation shining out into the ocean would be reflected back onto the dock, or into adjacent boats, where people might be exposed. None of us could calculate with confidence just how much radiation would possibly be reflected, and Rickover didn't want to leave any room for uncertainties. So he decided we should surround the full-scale hull of the submarine prototype plant with water, to simulate the actual submarine; then we could make the necessary radiation measurements.

The possibility that a submarine could be detected by its radioactive wake was also considered. But it was soon determined that any radioactivity induced in the seawater by the reactor would be completely trivial and undetectable compared with the considerable natural radioactivity of the sea itself.

Other locations around the shield were more controversial; for example, the passageway through the reactor compartment. Although you couldn't enter the shielded machinery compartment during operation, you still had to be able to get from one end of the ship to the other, so we built a shielded passageway through the reactor compartment. On the *Nautilus*, we stuffed everything into the lower half of the compartment and put a shield deck on top of it (Figure 6.4). This required us to have a horizontal boiler, with the steam chamber above and risers penetrating the shield, which was not an ideal arrangement. For subsequent ships, Bob Panoff, Rickover's Project Officer for Submarines, pushed a design he and system designers Milt Shaw and Howard Marks had been working on, consisting of vertical boilers and a shielded passageway, or tunnel, (Figure 6.5).

These three old war-horses, along with Karl Swenson, were involved in all the propulsion plant mechanical systems design and layout arguments. Panoff had worked for the Navy on non-nuclear submarine upgrade designs right after World War II, before reporting to Rickover. He continued to lead the various submarine design projects during his 15 years at Naval Reactors. Marks and Swenson had also worked in the Navy's Bureau of Ships before coming to Naval Reactors, and moved between submarines and surface ships during their careers. Shaw moved to the aircraft carrier project that followed *Nautilus*, and then went on to play a key role in the civilian nuclear power station project described in Chapter 8. They were all "engineers' engineers," pouring their passion into working closely and non-competitively with other engineers to coax nature into letting us create "things that never were."

The tunnel design gave a better machinery layout, but created some new shielding issues. "The compartment is only twenty feet long and there are no watch-stations there," argued Panoff. "People will walk right through; they're not going to hang out there."

"But the shielding in the tunnel gives us weight high up in the boat," replied Rickover. "That will decrease its stability. And you've now got radiation leaking out of the upper half of the hull, which is more of a threat to people on the dock."

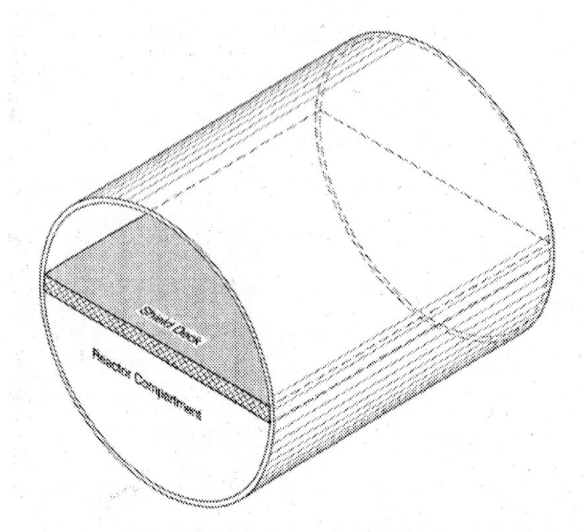

Figure 6.4 Shield Deck Design (*Hoke*)

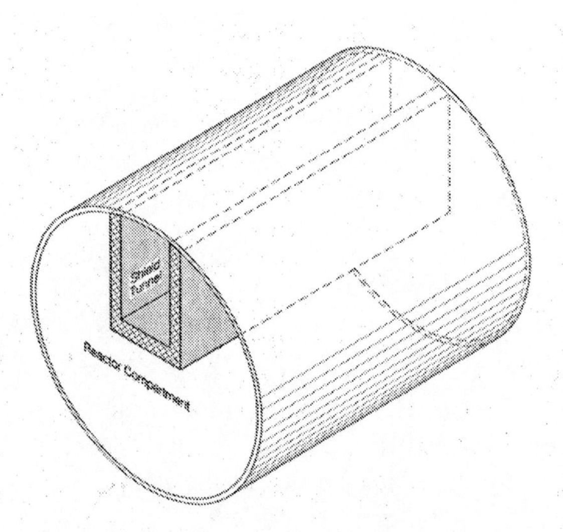

Figure 6.5 Shield Tunnel Design (*Hoke*)

"The shield wouldn't weigh so much if you didn't insist on shielding the tunnel as if it were a damn rest home. The radiation from the top of the steam generators probably won't be much higher than the radiation streaming through all the risers in the old shield deck." And so it went.

We finally settled all these issues, threw in some safety factor, and measured radiation around the operating plant. There were some problems that made us sweat, but we were able to work them out. We found that the shield was conservatively designed and built. And we proved that we knew enough to design and build shielding for the future, which was equally important.

Shield Materials

Before any serious efforts of design could be carried out, we had to decide what materials we were going to use for shielding. There were plenty of materials to pick from, each of which had special advantages and special disadvantages. Some people were pushing metal hydrides—exotic, unstable materials whose main feature is they contain a great deal of hydrogen. Hydrogen atoms are about the same mass as neutrons, and that turns out to be important.

If you want to stop a fast-moving object (like a neutron), another object the same mass is the most effective. Picture a fast-moving marble. If it hits a tiny BB shot, it will knock it aside and hardly slow down. If it hits a billiard ball, it will ricochet off and keep going. But another marble can stop it cold, if it hits dead-on. So hydrogen provides a marble-to-marble collision for neutrons and is thus very effective at slowing them down. Once slowed down, they're easier to stop.

But then we find another problem. Many atoms absorb slow neutrons effectively but give off a high-energy gamma ray when they do, which creates a new source of radiation to be shielded against. So another group of exotic materials was proposed, the rare earth metals. These are fourteen of the original 92 elements we all learned about in high school, but we never heard much about them because they're rare. They lie between lanthanum (number 57) and hafnium (number 72) on the Periodic

Table of the Elements, and they have names that sound as if they belong on Star Trek, names such as gadolinium, dysprosium, lutetium, ytterbium, promethium and praseodymium. Many of these elements are good at capturing slow neutrons, but not much is known about their properties, and nobody has built anything significant out of any of them.

Many other esoteric materials were suggested. One scientist kept insisting that pure gold was an excellent shield for gamma rays, and that tons of it were available from Fort Knox, not being used for anything else. When his technical arguments were rebuffed, he cried loudly that Captain Rickover was not willing to make the best technical decisions, being swayed by political considerations. Rickover's *ex tempore* response was a classic: "If you guys advised me that the best possible neutron shielding was white, Protestant babies (very hydrogenous), and that they were just lying around, doing no good, I'd still not use them, for political, not technical, reasons."

We kept trying to calculate just how much we would gain by using such fancy, special-purpose materials as these. We concluded that although they appeared very advantageous in simplified calculations, by the time you got into a real power plant, with structural materials, insulated piping, and penetrations for control wiring and plumbing, the actual advantage of the fancy materials might become very small.

Furthermore, the problems of working with such unknowns would certainly not fade away correspondingly. The same tended to hold true for some of the sophisticated methods of calculation. Each of these questions, and others yet unknown, had the potential to become endless research programs for theoretical physicists, experimentalists, heat transfer and structural engineers and materials specialists.

Just before joining the Naval Reactors Program in 1949, I had been working at the Atomic Energy Commission's Oak Ridge National Laboratory with a physicist named Everitt Blizard and an engineer named Charles Clifford, who together had developed a very clever and practical way to measure the effectiveness of various shielding materials. They removed a two-foot square block of shielding from the old graphite reactor and built a water tank outside the shield (Figure 6.6).

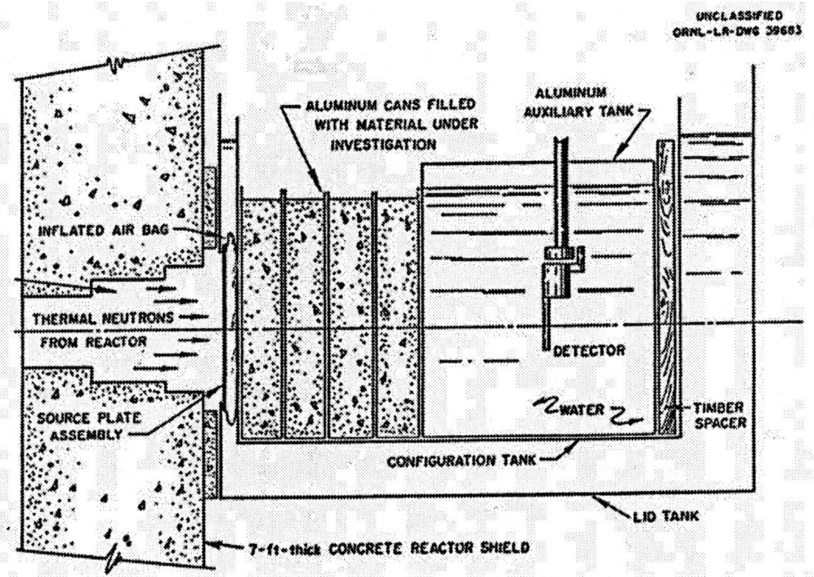

Figure 6.6 Shield test facility at Oak Ridge (*DOE*)

Inside of the tank, they placed a *source plate* of enriched uranium to convert the slow neutrons streaming out of the reactor to a fresh source of fission neutrons. They then placed slabs of various shielding materials in the tank and measured how effective these materials were at shielding the neutrons and gamma rays coming from the new source. It was a simple and elegant device and was instrumental in getting much of the data we needed to design the first radiation shields.

The simplest materials to measure in the tank were various combinations of iron and lead sheets interspersed with water. In the right order and thicknesses, these turned out to make very good shields. Water has lots of hydrogen in it, which is important for neutron shielding. But very high-speed neutrons, coming right out of the reactor, have to hit a lot of hydrogen atoms, or hit them just right, to slow down enough for easy capture. This is where iron turns out to be useful. When hit by neutrons, iron atoms act as if they were made of putty. The neutrons don't just bounce elastically, in billiard ball fashion. The iron actually absorbs some of the collision energy internally, and slows down the neutrons as much as if they had had several elastic collisions with hydrogen atoms. So we filled the tank with water, put in some iron sheets to knock some of the initial speed off the neutrons, then the water could finish the job. A typical set of measurements we took is shown in Figure 6.8.

Note that we are dealing with a very wide range of numbers here. Each big square in the chart is another factor of ten. The total range in the submarine was from about 10^{13} neutrons per square centimeter leaving the reactor each second (a centimeter is about 4/10 of an inch) to less than 10 neutrons per square centimeter leaving the shield. That's how much the shield had to reduce the neutron level.

As I mentioned earlier, it is usually important to ensure that the vertical scale of a graph goes in equal increments from zero to the maximum value. But if we did that here, Figure 6.8 would look like Figure 6.7 below. (Each step up in Figure 6.8 is another ten-fold increase.)

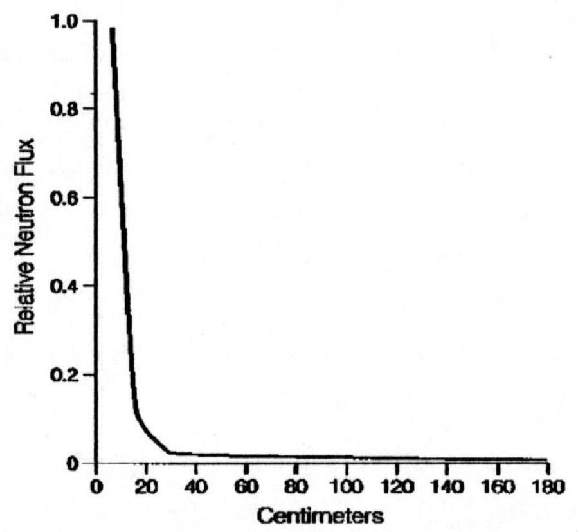

Figure 6.7 Shield data as linear graph (*Author*)

The neutron level decreases tenfold in each ten centimeters. So beyond about 30 cm, points are below 1 percent of full-scale and cannot be read. When you look at any graphed data, be sure to notice how the vertical and horizontal scales are marked. Does each division equal a fixed *amount* or a fixed *percentage*?

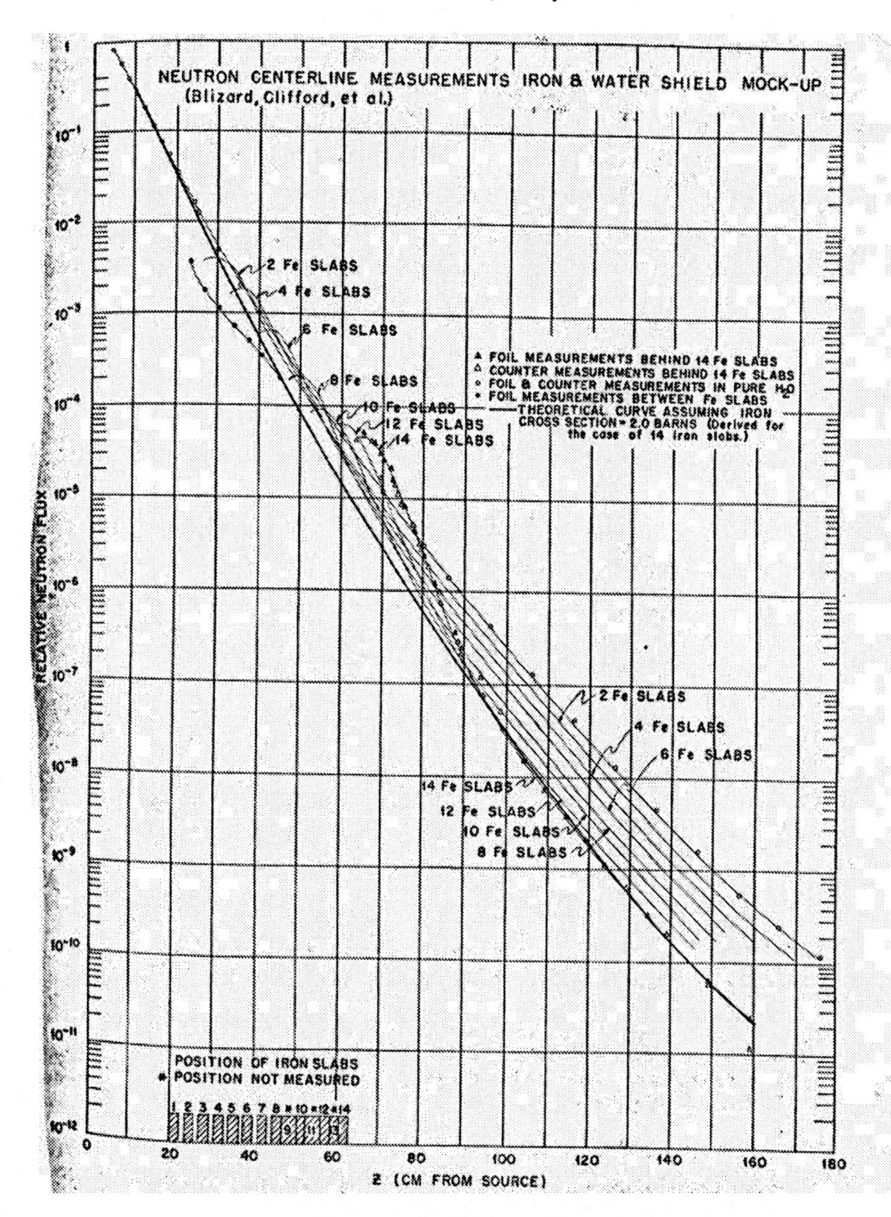

Figure 6.8 Shielding data for submarine shield, measured at Oak Ridge (*Author*)

The biggest practical problem we had was streaming of radiation up the insulation around the reactor vessel and its piping and penetrations. Insulation is inherently poor at stopping radiation, so patching around the "leakers" continued to be a problem until we had built a number of plants. We used chunks of polyethylene, which has more hydrogen than water, for this purpose. Noting that this streaming problem existed with all manner of solid materials but not with water, Rickover celebrated our decision to use water wherever practicable by noting solemnly "Water has no cracks."

The Reactor Shielding Design Manual

As an aid to his burgeoning program, with new laboratories and shipyards springing up like mushrooms after a rainstorm, Rickover reactivated the process he started in Oak Ridge of having his people prepare definitive textbooks on key aspects of the technology. I got shielding. I lined up the top shielding people at the Bettis and Knolls Atomic Power Laboratory (KAPL) laboratories and from the Electric Boat and Newport News Shipbuilding companies. We agreed on topic assignments and deadlines. But when the chapters came in, there were many overlaps and contradictions.

Streaming of gammas through cracks and through flaws and voids was covered analytically, but there was no comparable discussion of the even more troublesome neutron-streaming problem. So one of the authors wrote that up, quite elegantly too. But I asked, "I've never seen any analyses like that used for neutrons on our work. Does anybody actually do it that way?" They didn't. It turned out that, unlike lead, you could hold a thick sheet of polyethylene up to the light and see any physical flaws. They usually occurred only at the edges, and so you just trimmed them off. We dropped the analytical write-up.

We found we just couldn't resolve the many problems of overlaps, omissions and contradictions by phone and correspondence. So I talked General Dynamics (which had bought up Electric Boat Company) into letting us use a conference room in their Washington offices, and for a week of long days, we hammered out a finished product. Instead of chapters by individuals, each chapter now had many authors; ultimately, the book had ten authors, eleven contributors and one editor.

The manual was published in 1956 and received twenty or thirty reviews, mostly quite favorable. After several printings by the Government Printing Office, both McGraw-Hill and Van Nostrand published their own editions, several printings, and it was republished abroad in other languages. It is still in use today. The 100,000th graduate of the Naval Reactors Training Schools recently testified that the students all keep their copies handy. In 1986, I was given a "Lifetime Contribution Award, henceforth known as the Rockwell Award," by the American Nuclear Society, citing the Manual as the distinguishing contribution. Accepting the award somewhat sheepishly, I noted that the book was the product of many people, and I gave the check that accompanied the award to the Maryland University chapter of the American Nuclear Society, which was doing excellent work.

Years after the book was published, Henry Stone of GE's KAPL laboratory and a shielding engineer in his early days, called me with a wonderful story. He had gone to Japan to try to open that new market to GE's emerging commercial nuclear power program. The Japanese official was asserting that Japan did not need any assistance from American salesmen. Henry tried to convince the official that he was indeed a working engineer, but he was not getting through. Then he spotted a copy of the Shielding Manual on the official's desk. He showed the official his name on the title page. "Ah," said the official, his eyes lighting up. "Perhaps we can then find some things to talk about."

Developing a Theoretical Base

No scientist ever likes to proceed on the basis of experiments alone. There has to be some theoretical understanding of the processes going on. For one thing, our data came from the radiation impinging on the Oak Ridge shield test facility, which was much less than the submarine shield would experience. So we were designing several factors of ten beyond the experimental data. We needed some theoretical basis for doing this.

For both fast neutrons and gammas, the theorizers started by assuming the radiation was undergoing a series of billiard ball-type collisions in the shield, losing a little energy each time. Although gamma rays, like x rays or light rays, can be thought of as waves under some circumstances, they also act like particles when it comes to atomic collisions. They have an effective mass and a velocity, and their interaction with matter can be quite accurately calculated as if they were tiny billiard balls.

This is an easy calculation in concept, but because of the number of calculations required to simulate enough collisions to reduce the radiation a million-million times, it strained the capabilities of the crude computers available in the early 1950s. We called these "Monte Carlo calculations," because each radiation particle was undergoing a game of chance as to how it would fare on each collision and whether it would be one of the lucky few to emerge from the other side of the shield.

In addition to the straightforward collision process, there were various secondary events going on as well. Fast neutrons suffered inelastic collisions with iron thereby generating gammas. (Through such emissions, the iron shed the excess energy it absorbed from the neutron.) These gammas had to be shielded against. Even the slow neutrons created some gammas when they were absorbed, and the amount and energy of these gammas depended on what material happened to absorb the neutrons.

There were other, less important, secondary nuclear reactions going on that had to be identified, and the number and location of such events and the energy of the radiation they produced had to be calculated. The amount of heat generated by these processes was also a factor to be considered in the inner regions of the shield where the radiation levels were high. Once methods had been worked out for estimating all these things, we had to calculate what arrangement of the iron, lead and water layers would give us the best shield.

This configuration was then mocked up in the Oak Ridge facility, and measurements were compared with the calculated predictions. It was a back-and-forth procedure.

Completing the Design

The radiation from the reactor coolant system was nearly all gammas, so we shielded it with several inches of lead. The reactor itself was more complicated. We put steel plates inside of the reactor shield to initially slow the fast neutrons and followed this with water. Coolant pipes, electrical wiring and other penetrations through the shield tank had to be insulated, and this insulation created potential streaming paths right through the shield. Therefore, we inserted various steps along the way to break up the path, like the zigzags in an old Chinese bridge that are intended to make it difficult for demons to follow.

This question of radiation streaming was one of our biggest uncertainties. It was difficult to measure at Oak Ridge, and ambiguous results were obtained in measurements at atomic accelerators at MIT and elsewhere. Around the forest of control rods that penetrated the reactor *lid,* we could not use water for shielding; we relied instead on fabricated pieces of polyethylene plastic, which has about as much hydrogen in it as water has.

There is no good way to test a reactor plant shield except by running the plant and measuring what comes out. So as the time grew near to start up the plant, we had no real evidence that the shield would prove adequate. Since the submarine itself was well along in construction at the shipyard, any significant deficiency in the shield could deal a crippling blow to the whole project. You can't just throw a few more tons into a submarine at the last moment. So there were a few sleepless nights for many of us as we waited for the chance to test out this unprecedented design.

By the spring of 1953, we had built at the Idaho submarine prototype plant a complete, full-scale shield of water, iron, lead and a little polyethylene plastic. "We" includes the Electric Boat Division of General Dynamics (the shipbuilder) and designed by John Taylor, Kal Shure, Tony Foderaro, Fred Obenshain and others at the Bettis lab, and Ed Czapek and colleagues at Electric Boat. These were the people who developed the calculational techniques and computer programs to permit us to build the plant shielding, armed only with the Oak Ridge data and knowledge of how gamma rays and neutrons interact with materials. They are the unsung heroes of radiation protection.

The plant would be starting up soon, so we assembled a team to make the field tests, starting at low power and building up to full design power.

The Submarine in the Idaho Desert

The last place you might look for a sailor is the high desert country in the Valley of Lost Rivers on the Snake River plain near Arco, Idaho. The snow-capped Rocky Mountains look down on the barren desert, but it is no resort area. There is no Paradise Valley here, no Crystal Lake, no Magic Forest. Instead, the signs read Craters of the Moon, Blizzard Mountain and Cinder Cone. During World War II, the Navy had an ordnance testing range here, and the locals are used to steering clear of it. The U.S. Atomic Energy Commission had fenced off 439,000 acres—half the size of Rhode Island—and called it the National Reactor Test Station. Many of the KEEP OUT signs and the armed guards remain.

As you entered the building, you saw sailors clustered around McGaraghan Sea, which turned out to be a 185,000-gallon tank of water named after Commander Jack J. McGaraghan, USN, of Eureka, California, the naval construction officer in charge here. This was the world's first practical-sized application of the controlled use of atomic energy, a full-sized submarine propulsion plant, fitted into a submarine hull. It was nestled into McGaraghan Sea to simulate the submarine *Nautilus*, whose hull was already under construction 2,000 miles away at Groton, Connecticut.

A submarine is certainly the most difficult application for naval nuclear propulsion, and the Idaho desert would seem to be the hardest place to build a submarine, but there was indeed method in this madness. Submariners are used to staking their very lives on the complete reliability of others. Any one crew member acting carelessly could sink the ship. This heritage, plus the monastic isolation of the operators of the prototype plant, made it ideal for imposing the new kind of discipline necessary to operate a nuclear power plant. This mindset was easier to implant into these submariners than if we had had to start with the traditionally more relaxed surface Navy or the civilian operators. After the statutory Advisory Committee on Reactor Safeguards witnessed a series of drills aboard a pre-nuclear submarine—collision, fire, battle damage—they could under-stand the wisdom of this choice.

The submariners at this site were not typical sailors. Most of them were married, and they had to forfeit the sea duty bonus of about one-third of their regular pay to be here, yet they volunteered. Adding to the misery, the Navy did not provide living quarters, and the sailors had to scramble over the small towns in the vicinity to locate housing. They were expected to work twelve hours or more a day, plus a minimum of two hours commute time, six or even seven days a week. Yet, despite a normal amount of griping, they were generally quick to count themselves lucky. "There's only one chance to be first," was a common remark.

"Big Jack" McGaraghan was just the kind of guy this desert out-post needed. He knew how to get a job done, he could talk with Rickover without losing his cool, and he seemed to be crawling all over the job at every hour of every day. Making the 70-mile commute to the site on a cold pre-dawn winter's day, he would point out to me the graceful herds of antelope, the jack-rabbits lined up literally shoulder-to-shoulder along the steam pipes to keep warm, the occasional lone elk silhouetted against the moonlight, and the ever-present tumble-weeds. He would plug his car's block-heater into one of the outlets that faced each parking space to protect engines from the minus 35° cold, and start another 16 to 18-hour day.

Running the Shielding Tests

With full support from Captain Rickover and Jack Kyger, I was determined to accomplish a number of things with these tests. First, we brought in personnel from Oak Ridge who had been measuring the nuclear characteristics on which the shield design was based. Second, we also brought in Henry Stone and colleagues from the Knolls Atomic Power Laboratory (KAPL), which was General Electric's counterpart to the Westinghouse Bettis Lab.

Figure 6.9 Full-scale prototype submarine plant, hull and shield tank (*Westinghouse*)

KAPL was working on the sodium-cooled reactor design and was in the process of designing the shield for that plant. They had expertise, both experimental and theoretical, to contribute, and they would also learn a great deal for their own project by being personally involved in a full-scale shield test program. I insisted that we draw up maps of the plant, and at every point where radiation measurements were to be taken, calculated values would be written in ahead of time. This caused quite a howl of protest because some of the locations had such complicated structure that we could not make very good calculations. "That's why we're running the tests," they objected. "After we've made the measurements, then we'll know. If we knew how to calculate all these points, we wouldn't have to run the tests."

Their arguments certainly sounded reasonable, particularly when made very rapidly and very loudly. I was convinced though, that unless we predicted and wrote down specific numbers in advance, we would rationalize afterwards that all the measurements were about what one would expect, and we would learn little from them. It is from the surprises that we learn. So we prepared dozens of layout sheets showing at each location the expected values for fast neutrons, slow neutrons, and gamma rays of various energies, as measured by each of the instruments we planned to use. And in fact, this proved to be quite useful during the test program. In one location, for example, the slow neutron levels were coming in markedly higher than calculated. It became clear that this was more than *experimental error*. Something was wrong. Then we discovered we had inadvertently used indium foils rather than gold as neutron detectors. Each was scheduled to be used, but in different places. Once this was recognized, the numbers fell back into place. If we had not discovered this on the scene, it would have been a major project to figure it out back at the lab in Pittsburgh.

In other cases, we found some unreasonably high streaming of radiation, which turned out to be an installation problem: the pieces of polyethylene were not tightly fitted together. (This was minor at the prototype, but the same problem developed more severely on the ship, and on some subsequent ships.) We might not have realized these readings were high if we did not have the pre-calculations with which to compare them.

I found there is nothing quite as educational as taking field measurements yourself. It just isn't the same as making measurements or calculations in the lab or the office. I was measuring gamma radiation with a hand-held meter, from right against the forward reactor compartment shield and moving away from it toward the living quarters. As I slowly backed away from the reactor, the radiation level increased. *How can that be?* I puzzled.

After I checked it a few times, John Taylor joined me. "What's going on here, John?" I asked. He just grinned and handed me a copy of the appropriate test sheet. The calculated numbers did the same thing: they increased as you moved back from the shield, then quickly dropped off as you got still farther away. John showed me why this had to be.

The two biggest radiation sources at the forward bulkhead were the two big steam generators, one at each side (Figure 6.10). So when you stood close to the shield wall, the radiation from the steam generators at the sides had to reach you by taking a long slanting path through the thick shield, and was thereby reduced. Backing away, radiation could reach you by traveling less obliquely, on a shorter path through the shield. Beyond a few feet away, this effect was offset by the growing distance, and the usual decrease of radiation with distance took over.

The Results

The shield proved to be conservative enough to provide some margin for changes, yet close enough to make us feel confident in our methods of calculation. The only significant problem was neutron streaming around the control rod forest, and much of that turned out to result from difficulties in getting the odd-shaped polyethylene blocks fitted into place. As previously noted, radiation streaming up the thermal insulation was also troublesome. But these problems were fixed, and the calculational methods were well validated.

We slept better after that.

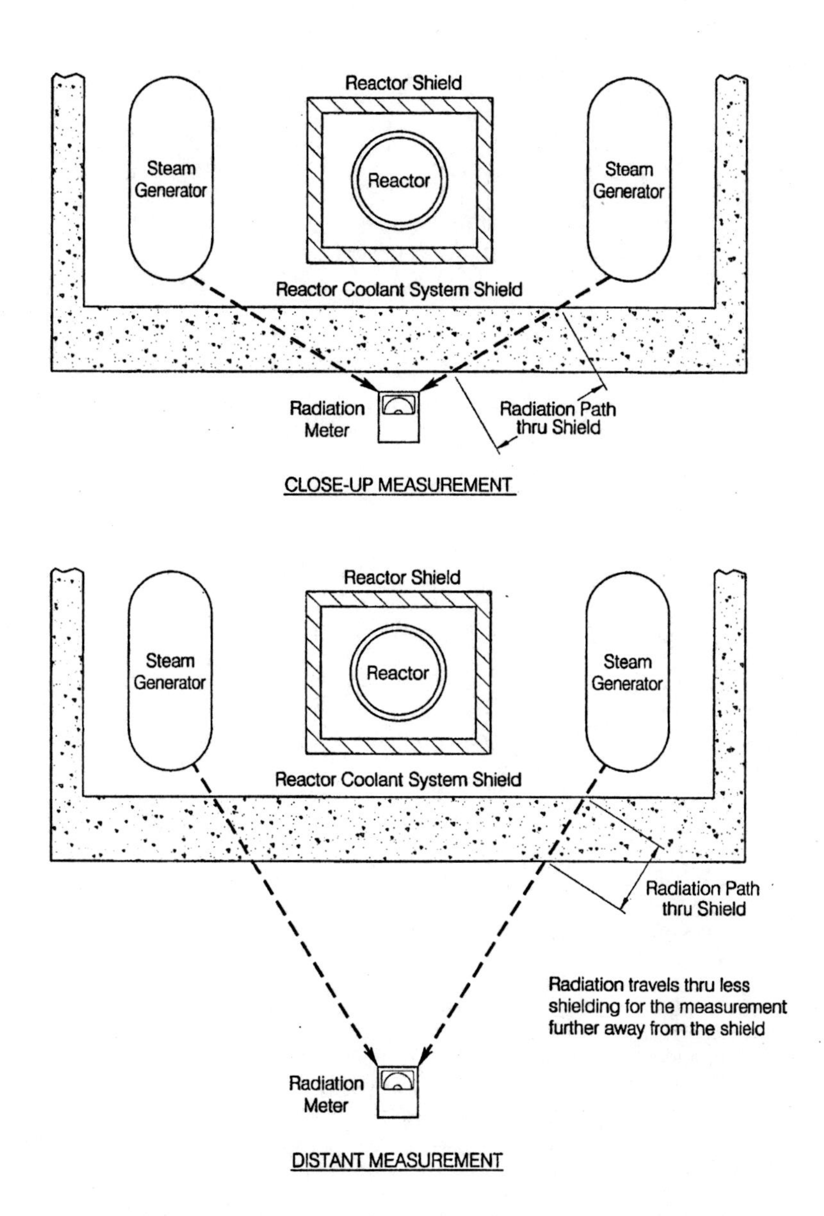

Figure 6.10 Radiation at forward bulkhead (*Hoke*)

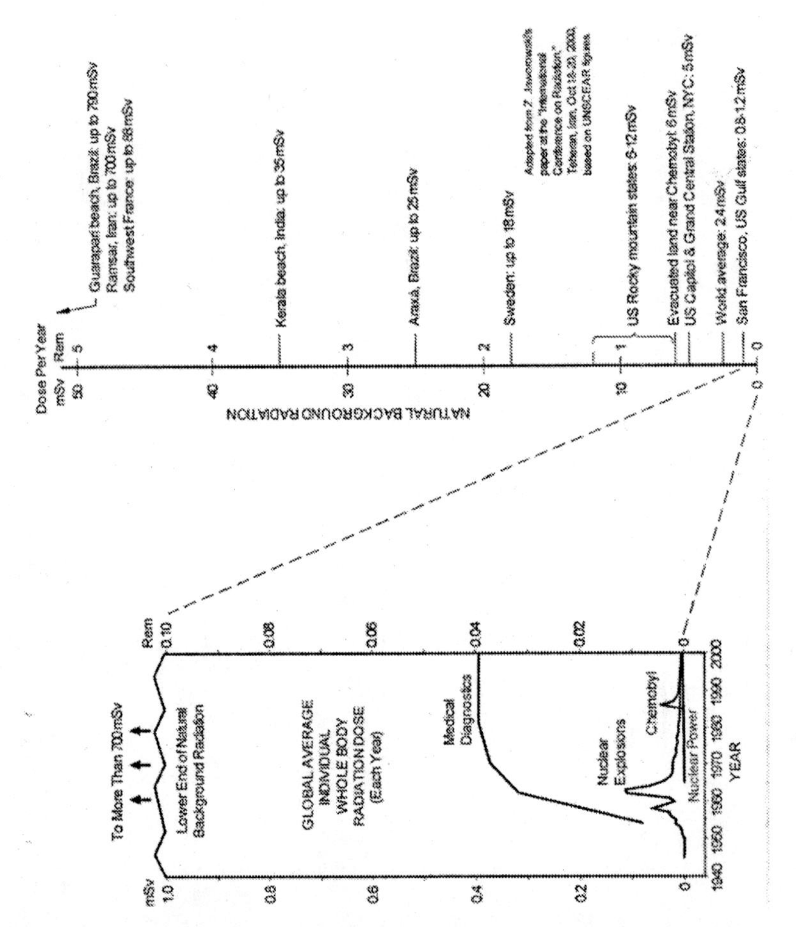

Figure 7.1 Sources of Radiation: The figure on the LEFT shows the average radiation dose we get each year from various sources, in Rem, the American unit, and in the metric unit, mSv. The radiation from nuclear power and its associated operations (including the Three Mile Island meltdown) hardly shows up. Fallout from the reactor accident at Chernobyl peaked, then declined. Fallout from testing nuclear weapons was larger, but it has also subsided. Medical diagnostics: dental and other x rays, radioisotope tests, CATscans (not counting the large radiation doses given to burn out tumors) are the largest. At the very top, we enter the lowest levels of natural radiation background. The highest natural background levels would be several hundred feet off the top of the paper!

So on the RIGHT, we have a reduced scale, going from 0 to 50 mSv per year. The numbers on the first chart are now all squeezed in between 0 and 1. The world average is 2.4. Some rooms in the US Capitol building are over 5. Places in Sweden are 18, parts of southwestern France are 88. And there are places in Iran that are over 700! These are places where generations of people have lived healthy long lives. USEPA says 0.04 mSv per year is too high!

I do not hesitate to say that this is
the greatest scientific scandal of the century
— Prof. Gunnar Walinder, Former Chair, Swedish Radiobiology Society

LNT: The Linear No-Threshold model, a postulated relationship between the amount of ionizing radiation striking a person and the resulting detrimental health effects. The model says that the damage is linearly proportional to the radiation dose down to zero dose, and that no dose is small enough to be harmless. This model is the basis for all radiation protection regulations, standards and procedures. It was created for administrative simplicity and is flatly contradicted by a vast body of credible scientific data and theory.

7. The Great LNT Scandal

While we were designing, building and testing the first reactor shielding installations, we didn't think much about the biological process of how radiation interacts with living organisms. We didn't have to. There was plenty of evidence that *high* levels of radiation could be harmful. And there was wide agreement that the permissible levels set by radiation protection standards were conservative. In fact, we know more about the biological effects of radiation than about most other biological hazards we face, such as the toxicity of fumes from kitchen grills and industrial smokestacks and trace toxicants in our food and drinking water. We found we could meet these conservative standards. There was no reason for confusion or conflict on the subject. But that happy situation did not last.

After I left Naval Reactors in 1964 and began to see a broader view of the nuclear enterprise, I became aware of a situation that had been building up for some time. A number of critics were questioning the adequacy of the protection standards. Much of this questioning was simply anti-nuclear rhetoric with little attempt to justify it scientifically. But subtler minds began to build a scientific rationale. They argued: Suppose an individual shows no immediate harm from a radiation dose, but then decades later comes down with cancer. How do we know that the cancer was not caused or abetted by the earlier radiation dose? Since 30 to 40%

of all persons get cancer at some time in their lives, this question is a troublesome one.

The Effects of Radiation on People

The most important question concerning the effects of radiation on people is: Is radiation always harmful, no matter how little we get? To answer this, let us look first at how radiation interacts with the body, quoting from Sheldon Novick's anti-nuclear book *The Careless Atom:*

> When one of these particles or rays goes crashing through some material, it collides violently with atoms and molecules along the way… In the delicately balanced economy of the cell, this sudden disruption can be disastrous. The individual cell may die; it may recover. But if it does recover … after the passage of weeks, months or years, it may begin to proliferate wildly in the uncontrolled growth we call cancer. (page 105)

That certainly sounds dangerous. And twice as much radiation will affect twice as many cells. In view of this, how can anyone possibly argue that "a little more radiation won't hurt"?

The answer lies in the numbers. Remember Lucy in the Charles Schulz "Peanuts" comic strip? She wanted to get a good look at the stars, and in order to get as close as possible, she stood up on a little chair. Her logic is unassailable:

1. You can see things better if you get closer.

2. Standing on a chair gets you closer.

This is foolish only when you consider the numbers. Lucy could climb up on a table, or to the top of a ladder, but her distance from the stars would not change significantly. The distance to the stars is so great, and the length of the ladder is so short in comparison as to be insignificant.

How does this apply to radiation? First, we must realize that the body sloughs off billions of dead cells every day in its continuous process of renewal. We don't assume that a pine tree is dying just because we see lots of dead needles under it. We know that dead needles are a natural by-product of normal pine tree growth. There are many more cells in a human body than needles on a pine tree. And 98% of the atoms that make up these cells are replaced each year by atoms from the food we eat and the air we breathe. So before we worry about those cells damaged by radiation, we should ask how the number killed by radiation compares with those routinely killed in this natural process of metabolism and regeneration.

The scientific evidence is clear: for every cell killed by natural background radiation, millions are killed by this natural process of bodily renewal. But what about the damaged cells *not* killed? Isn't that where cancer comes from? No. The fact is that only one in ten million human cancers is caused by radiation, natural or man-made. The odds against one of those damaged cells leading to cancer is estimated to be about one in 10^{24}. (That's a one with twenty-four zeroes after it!)

The LNT Model

To be conservative, an *administrative* decision was made in the early days of nuclear energy to assume that at low doses, radiation continues to be harmful in proportion to the dose, all the way down to zero. This is shown by the dashed straight line labeled "LNT" (for "Linear No Threshold") in Figure 7.2. This is the origin of the idea that "no amount of radiation is harmless." There has never been any scientific basis for this assumption. It was mentioned by the International Commission on Radiological Protection (ICRP) in 1960 and recommended by ICRP in 1972. But this concept leads to the silly notion of *collective dose*: if 1,000 rem can kill one person, then one rem to each of 1,000 people is supposed to cause one fatality (somewhere) and so will 1 millirem to each of a million people. We don't make that sort of assumption for any other substance, and its use in radiation protection is scientifically indefensible. We know that if no one gets a harmful dose, then no one is harmed.

Even the data on high-level radiation doses have some conservative biases. The laboratory data on irradiated mice are invalidated to some extent by the fact that mice are known to be more sensitive, and differently sensitive, to radiation than humans. And we are beginning to recognize that the inbred mice and rats used for laboratory tests have vulnerable immune systems. Tests run on wild chipmunks show much less damage from radiation than tests on laboratory animals. Moreover, there is recent research showing that laboratory animals fed *ad lib*, that is rats and mice allowed to eat all they want, become obese and further weakened. Their life spans and their incidence of cancer are dramatically higher than for animals fed in a controlled manner.

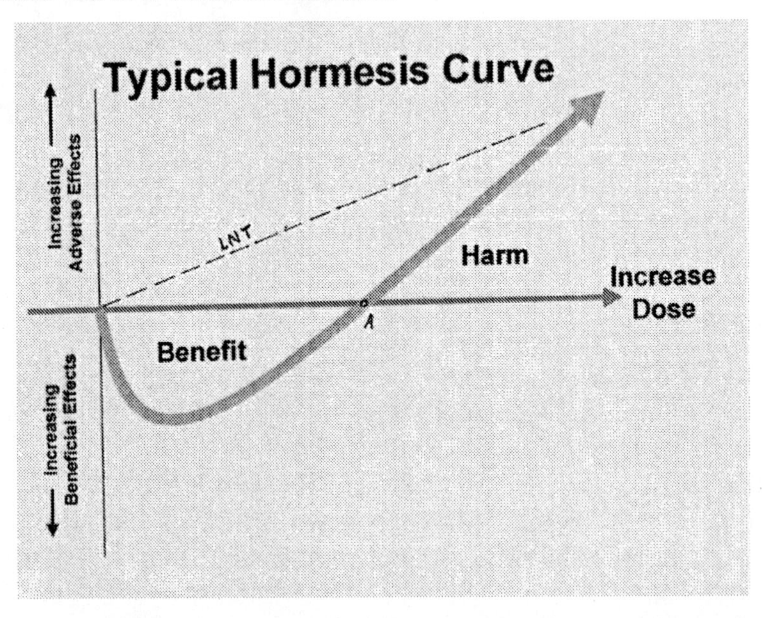

Figure 7.2 Biological Effects on Typical "Toxins" (Including Radiation)

In addition, radiation received rapidly from an atomic bomb or from irradiation tests in a laboratory, is much more toxic than radiation received more slowly, allowing time for the body to heal. Dr. Lauriston S. Taylor, one of the great radiation protection pioneers, notes that if people are exposed to 350 to 400 rems in a short burst, about half of them would die within 30 days. "By contrast," he writes, "the same dose administered uniformly over a year's time could pass unnoticed by most exposed persons." This should not surprise us; we know that a large bottle of pills (such as aspirin) taken one a day might be beneficial, whereas gulping them all down at once could kill us.

No Hiding Place Down Here

We can't look at how things would be without any radiation, because that situation does not exist—not on this earth, or in outer space, as far as we've explored it. There is no place we can go to get away from radiation. God's good green earth is in fact a naturally radioactive atomic waste dump, composed of the waste products of all the radioactive processes that produced the sun and light the stars. Our soil is naturally radioactive, and so are the oceans. The center of the earth is still molten because of the heat generated by this natural radioactivity, without which our planet would be cold and lifeless. Cosmic rays from beyond the galaxy bombard us from above; naturally radioactive potassium, carbon, rubidium, tritium, uranium, thorium, and their many radioactive decay products irradiate us from below. Our very blood and bones are radioactive, with half-lives up to a billion years (that is, it takes a billion years for their radioactivity to decay to half, and another billion years to decay to half of that). This situation began before the arrival of humans on earth and has nothing to do with our nuclear power activities. In fact, there is evidence that radioactivity is essential to life as we know it. Figure 7.3 compares some of the sources of radiation that confront us in everyday living. It shows that even making pessimistic assumptions about nuclear reactors, the radiation originating with nuclear power is tiny compared with the radiation from natural sources.

Trying to Minimize Your Radiation Exposure

If you are really concerned about reducing the amount of radiation your body receives, you might try to find a place to live where the natural radiation level is lower. You'd have to avoid flying and skiing and stay away from Colorado; in each case, the thinner atmosphere lets in more cosmic rays from space. (At sea level, the air shields cosmic rays as effectively as twelve feet of concrete.) Also stay away from parts of New England where the granite soil contains a lot of uranium. And parts of Florida, which has phosphate rock (used to make fertilizer) that is also quite radioactive due to its uranium content. And the fertile Piedmont Plain section of America has a high radon level. And the water in Wisconsin has radium in it. Don't live in a stone or a brick house; these also emit natural radiation. And don't sleep with anyone; the natural radiation from another body (either sex) is yet another radiation source. But most of us will choose to live without such precautions, believing (correctly) that such low levels of radiation pose no real hazard.

Every day, over 300 billion of our body's cells are struck by radiation from these natural sources. Natural radiation causes 70 million DNA-damaging events in each of us every year. This is for a typical environment. There are places in the world where people have lived healthily for countless generations with natural radiation over a hundred times greater than other places, and they thus get correspondingly more initial cell damage. About 12 million Americans get more than 1,000 mrem per year to the lungs, and about 2 million of these get more than 2,000 mrem per year just from radon. But they show no harmful effects from this. On the contrary, detailed studies have shown that they generally live at least as long and are healthier than people who live in locations with much lower radiation levels.

Hormesis: the Beneficial Effects of Radiation

To many people, the idea that radiation could be good for you lies somewhere between the absurd and the insidious. It reminds them of the 1960s joke about the tobacco companies deciding to fight the Surgeon General's report with advertisements proclaiming: *Cancer is Good For You.* But in fact, there is solid scientific evidence that small quantities of radiation are beneficial—perhaps even necessary—to health.

This idea that toxic materials are beneficial at low doses is not new, nor is it confined to radiation. The proto-scientist Paracelsus stated it clearly in 1540: "Nothing is poisonous, but the dose makes it so." This principle is called *hormesis*, from the Greek word to stimulate. It refers to the fact that tolerable challenges to any organism stimulate the immune system and strengthen the organism. Any fight you win makes you stronger. Toxicologists E. J. Calabrese and E. A. Baldwin stated in the authoritative journal *Nature*: "The hormetic model is not an exception to the rule--it is the rule." (Feb 13, 2003) It would be anomalous if radiation behaved differently.

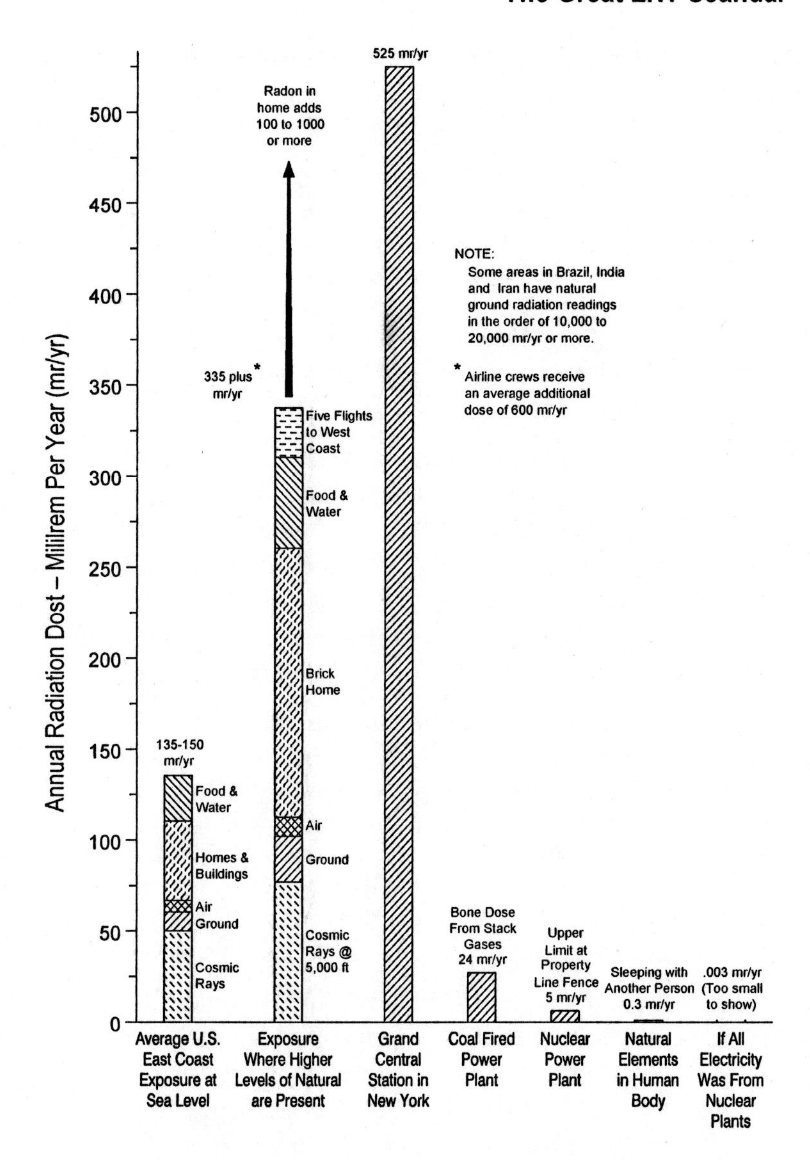

Figure 7.3 Some everyday radiation sources (*Hoke*)

Over forty years ago, Dr. Hugh Henry summarized for the *Journal of the American Medical Association* (vol. 176, p. 671, 1961) some of the Oak Ridge studies. His conclusions are clear:

> A significant and growing amount of experimental information indicates that the overall effects of chronic exposure (at low levels) are not harmful...The preponderance of data better supports the hypothesis that low chronic exposures result in an increased longevity than it supports the opposite hypothesis of decreased longevity... Increased vitality at low exposures to materials that are toxic at high exposures is a well-recognized phenomenon.

T. D. Luckey, Chairman Emeritus of the Department of Biochemistry of the University of Missouri School of Medicine, published a book on "Hormesis" (CRC Press, 1980) devoted entirely to the beneficial effects of low-level radiation, citing 1,269 research reports, and followed with another book (CRC Press, 1991) similarly titled, citing 1,018 references. The abstract of his recent summary of the situation (*Radiation Protection Management*, vol. 15, p. 19, 1997) states:

> Exposed nuclear workers and military observers of atmospheric atomic bomb tests with carefully selected control populations provide 13 million person-years of experience with low-dose radiation. These carefully monitored studies show conclusively that low doses of ionizing radiation reduce premature cancer mortality in humans. When person-years were used to obtain a weighted average, the cancer mortality rate of exposed persons was only 65.9% that of unexposed controls.

The solid curve in Figure 7.2 shows the biological effects of most toxic substances, such as lead, mercury, arsenic, copper, selenium, manganese, chromium, etc. Radiation seems to act the same way. Below zero on the damage scale there is negative damage—that is, benefit. Damage can be in the form of increased cancer incidence, decreased longevity, etc. From zero radiation (above background) to the point A, the body has a radiation deficiency and would benefit from more radiation. For radiation doses above A, there is damage—more and more damage as the dose increases. The natural radiation background is generally in the beneficial region, i.e., nearly all of us could benefit by more radiation.

A great deal of research shows that the beneficial effect of small doses results from the toxic material acting like a vaccine, stimulating the body's anti-mutagenic defenses. These defenses work to prevent damage, to repair damage, and to remove damaged cells from the system so they can't go on to become cancers. Laboratory tests show that low-dose radiation stimulates each and every one of these cancer-fighting defense processes. *This enhancement of the body's defense processes is not limited to the occasional cell damaged by radiation. It works on all cells, including the 10 million times larger number damaged by normal metabolism. So the net result is to decrease the number of persistent mutant cells that lead to cancer*.

The science is very clear on this. But at the time of my writing, this important fact has not yet been taken into account in setting radiation policy. I'll come back to that in a moment.

Selenium is a good example of a poison acting beneficially. It is considered highly toxic. Cattle, horses and sheep grazing in selenium-rich soil lose their hair, their appetites, become paralyzed, and die. Yet a selenium *deficiency* causes other problems, including an increased susceptibility to cancer and congestive heart failure. Serious illness and multiple deaths among grazing animals has been traced to a deficiency of selenium in the soil. The minimum intake recommended to maintain health is about one ten-thousandth of a gram per day. And a gram is only one twenty-eighth of an ounce. Yet somewhere between three and five ten-thousandths of a gram is considered an upper safe limit. Luckily, we have a much greater tolerable operating range.

During the past fifty years we have accumulated lots of data on low-level radiation. This work generally confirms the solid curve of Figure 7.2 and refutes the linear assumption (marked "LNT"). It shows there is a threshold—somewhere between 20 and 100 rem per year—below which there are no detrimental health effects from radiation, and below 20 rem the organism may actually suffer from a radiation deficiency. More research is needed to determine optimum radiation doses and dose rates, and to explore the variation with different kinds of cancer. But this is not the kind of work that can currently get funding from the radiation protection funding sources.

The conclusions stated above come from international teams of scientists and physicians studying: 1) Hiroshima and Nagasaki survivors; 2) occupational exposure among radiologists and atomic energy and weapons workers in the U.S., the U.K., Canada, and the former U.S.S.R.; 3) medical patients receiving radiation therapy; 4) persons who ingested radium during the days when radium was used to make luminescent dials; 5) miners working in uranium mines; 6) U.S and British troops participating in atomic bomb tests; 7) persons living in high-level natural radiation backgrounds; 8) and laboratory tests on plants and animals.

If current regulatory policy and practices were changed to reflect the scientific evidence—that small amounts of radiation can be beneficial rather than harmful—this would have a very significant impact on radiation protection and environmental clean-up planning for the immediate future and for the long-range. And it could significantly reduce much of the fear that surrounds the very word radiation in many people's minds.

Low-dose x rays have been used for nearly a century to treat local infection and avert the need for amputation. The radiation is too weak to kill the bacteria, but it stimulates the immune system to do the job. When sulfa and other "wonder drugs" were introduced, these became the treatment of choice, although they are much less effective. Clinical tests were run at Harvard in 1976 to successfully treat cancer with whole-body x rays, and Sakamoto and others in Japan researched the process in more detail and reported further successful treatments.

When a colleague of mine, E. J. Bauser, contracted an "incurable" cancer, he volunteered to take the treatment. His primary physician formally terminated their relationship warning that the radiation would kill him. The treatment gave substantial improvement with no detectable side effects, though as with chemotherapy further treatments were required as the condition returned. As of this writing, 11 years after his original "terminal" diagnosis, he has been unable to find a therapist willing to give further radiation treatments. He is told, "Chemo is the recommended treatment." But at age 85, he dreads the prospect of the debilitating effects of further chemotherapy.

Is Ionizing Radiation Essential to Life?

We've seen that low doses of radiation are beneficial to life. What if we were to reduce radiation levels to below the natural background? How would an organism respond to that? There were hints as to what the answer would be when active marine life was found near hot underwater jets, thriving deep in the sea, far below the reach of life-supporting sunlight. These areas are also high-radiation zones because of the natural radioactivity flushed up with the jets. Lacking sunlight, these flourishing biota may derive their energy from the radioactivity. But no money is available to investigate such an exciting possibility.

A number of experiments have been done with plants and mice and other animals that were not only shielded from external radiation, but in the case of mice, were fed special food whose radioactive potassium isotope was depleted, greatly reducing their bodies' natural radioactivity. The organisms looked normal but failed to function properly. When radiation was restored, they returned to normal functioning. Again, these studies have not been properly written up and followed up on since they contradict rather than support official LNT doctrine that all radiation must be harmful.

Radon: "The Silent Killer in Your Home"

High doses of radon are said to cause lung cancer. Therefore, based on the linear no-threshold (LNT) model, regulatory bodies claim that natural radon levels in homes present a risk of lung cancer. The BEIR-VI report, "Health Effects of Exposure to Radon," by a special committee of the National Research Council, concluded: "the estimated 15,400 to 21,800 deaths attributed to radon … constitute an important public-health problem." (I won't comment on their implication of three-figure precision in an estimated range that varies by over 40%.)

Let's look at the data. Figure 7.4 shows the number of lung cancer deaths at various levels of radon as reported by Dr. Bernard Cohen. The figure is based on actual radon measurements in homes and number of lung cancer deaths in counties in which over 90% of Americans live. This is the actual population to which radon regulations apply. The dashed line shows the LNT "prediction" of lung cancer deaths for this same population. The data make clear that, within the range of radon levels measured, the number of lung cancer deaths *decreases* as radon levels increase. It does not increase, as the EPA and the National Research Council reports claim.

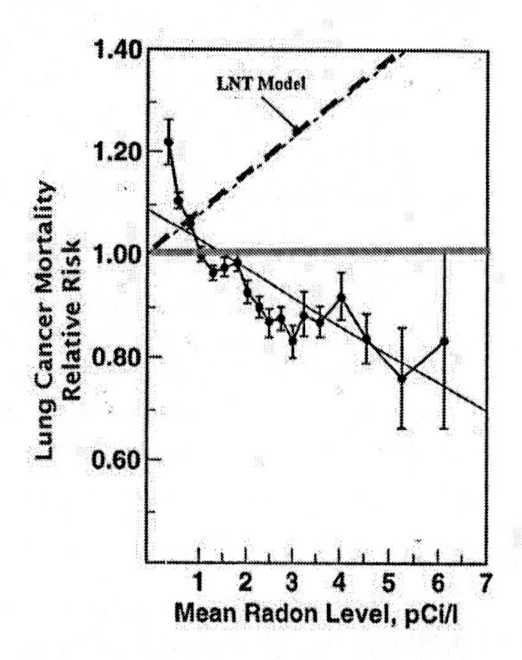

Figure 7.4 Cohen's lung cancer mortality data at various radon levels (*Cohen*)

Since this evidence differs from the LNT by twenty standard deviations, the policymakers have tried to ignore, obfuscate and disparage it. But their criticisms take the form of generic objections, which don't in fact apply to the actual case. For example, the BEIR-VI report on radon, on which the EPA regulations are based, relegates this evidence to its

Appendix G and doesn't even mention the many other studies that reach the same conclusion. Appendix G states:

> Potential confounding by smoking was addressed ... The potential for confounding by sociodemographic factors or their correlates was explored by stratification on levels of 54 variables. Confounding by geography was assessed by stratification, and the sensitivity of the findings to outliers was examined. There was a strong negative association between 1970–1979 lung cancer mortality and the county-average radon concentrations; the association could not be explained by confounding. In interpreting this finding, Cohen proposes that the negative association implies failure of the linear non-threshold theory.

Sounds pretty convincing, doesn't it? Let's see what the body of the report says. The Executive Summary Conclusions states: "The carcinogenicity of radon is convincingly documented through epidemiological studies of underground miners, all showing a markedly increased risk of lung cancer." But miners are exposed to diesel fumes, silica and other mineral dusts, as well as higher radon levels than found in homes. Certainly all high-radon home-dwellers do not "show a markedly increase risk of lung cancer." The Conclusion concedes, "most of the radon-related deaths among smokers would not have occurred if the victims had not smoked."

But what about Cohen's data on actual people living in homes? This is not even mentioned in the entire 14-page, single-spaced Executive Summary. Nor is it discussed in the main body of the report. Going back to the 61-page Appendix G, we find that "ecological studies" (the type that Cohen and others performed using average radon measurements and average lung cancer data) are dismissed as follows:

> We conclude that ecological studies are noninformative for estimating risks posed by exposure to indoor radon or for evaluating a potential threshold exposure below which radon progency exposure would not be harmful.

So on what does the BEIR committee base its conclusion that 15,400 to 21,800 Americans die each year from radon? (See Figure 7.5.) Do you find that more convincing than Cohen's data? I sure don't!

Earlier in the appendix, a critic of Cohen's data states: "Most of us would not be willing to discard a useful theory (i.e., the LNT premise) on the basis of such a test." This turns on its head the classical scientific method of Sir John Popper and Richard Feynman, which requires that "a theory, however elegant" must be abandoned if it is contradicted by a single immutable fact. And here we have not just a single fact, not just the mass of radiation data, but everywhere else we look—toxicology, vaccinations, sunshine, exercise therapy—all exhibit the biphasic response: harmful at high levels, beneficial at low. In the case of radiation protection, that principle seems to be repeatedly overlooked.

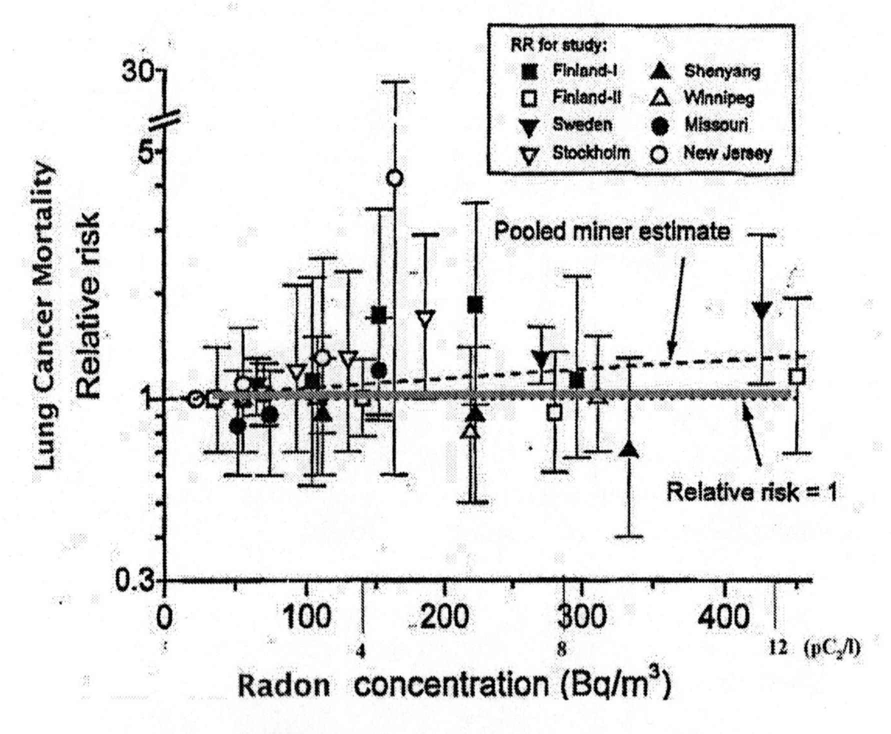

Figure 7.5 BEIR VI graph of lung cancer vs. radon levels. Note size of error bars (*National Academy Press*)

How Is Such a Discrepancy Maintained?

How do LNT advocates respond to the charge that the evidence does not support their premise? I am continually surprised at how little effort they make to state a scientific defense. For example, in the report NCRP-121, p.45, they state:

> *Few experimental studies*, and essentially *no human data,* can be said to prove or *even to provide direct support* for the concept ... The *best* that can be said is that *most* studies do not provide quantitative data that, with statistical significance, contradict the concept....It is *conceptually possible*, but with a *vanishingly small probability*, that any of these effects *could* result from the passage of a single charged particle, causing damage to DNA that *could* be expressed as a mutation or small deletion. It is a result of this type of reasoning that a linear nonthreshold dose-response relationship *cannot be excluded.* [Emphasis added.]

In June 2001, after six years of study, report NCRP-136 recommended continued use of LNT but conceded (page 6):

> It is important to note that the rates of cancer in most populations exposed to low-level radiation have not been found to be detectably increased, and that in most cases the rates have appeared to be decreased.

With such a weak case, you would think that it would not be possible to maintain such a discrepancy between science and policy for several decades—nearly two human generations. In controversies outside the nuclear field, there are people advocating tightening of safety standards and others arguing that excessive regulation is costly and wasteful. These two forces tend to be resolved by a middle-of-the-road solution that is tolerable to both sides. But for nuclear power there has been no force for moderation because all parties (except consumers) have profited from the fruits of radiophobia. So we have neither a personal nor an institutional constituency for radiation reform.

Researchers, policymakers and regulators draw their incomes and their reputations by continually studying a problem that is said to be dangerous and mysterious. Industry benefits from having lucrative projects to create large, complex safety systems and by "decontaminating" and "remediating" trivially-radioactive sites and equipment. Efforts to show that such measures are unnecessary are met with warnings, "The Government has been nice to us; we don't want to disparage them."

Leo Tolstoy said it well in his 1901 book, *What is Art?*:

> I know that most men, even those who are clever and capable of understanding the most difficult scientific, mathematical or philosophical problems, can seldom discern even the most obvious truth if it be such as obliges them to admit the falsity of conclusions they have formed perhaps with much difficulty—conclusions of which they are proud, which they have taught to others, and on which they have built their lives.

How Can Radiation Protection Policy Be Changed?

To challenge this situation, James B. Muckerheide, State Nuclear Engineer for Massachusetts, and Co-Director of the Center for Nuclear Technology and Society at Worcester Polytechnic Institute, set up in 1995, an international not-for-profit organization of independent experts on radiation science and public policy and called it Radiation, Science & Health, Inc. (RSH). Its stated mission is: "To document the scientific data that contradict the linear model;" and "To advocate for revision of radiation science policies."

RSH has collected, evaluated, excerpted and published on its website, **http://cnts.wpi.edu/rsh/docs**, evidence refuting the LNT from several thousand papers. Muckerheide recruited respected senior scientists from all over the world who were retired or otherwise able to resist pressure from the radiation protection community. He asked me to be a founding officer and director, and I gladly accepted.

Muckerheide had already started in 1994, arranging special sessions at the annual meetings of the American Nuclear Society (ANS), where scientists could present their research data refuting the LNT, answer questions, and discuss the implications. Seventeen such sessions were held, and the many papers presented were made part of the ANS

Proceedings, available to the scientific community at large. In addition, RSH provided speakers for dozens of other technical meetings worldwide. It was becoming increasingly difficult for policymakers to claim they were not aware of evidence that warranted changing the policy.

We started with our professional society, the American Nuclear Society, and tried for five years to get a simple position statement that low-dose radiation was not hazardous, and thus, the LNT should not be used to set radiation standards for the low-level radiation that is relevant to nuclear facilities. One would expect a scientific society to be reasonably free of political concerns. But one of the members kept expressing concern that such a statement might imply that the National Council on Radiation Protection and Measurements (NCRP) was not doing its job.

The statement was blocked, and the objector was later appointed to the NCRP. Another, expressing concern in more generic terms, was then appointed to the ICRP (the international equivalent). After it was too late to help shape public opinion on a number of important issues, a lengthy but fairly good compromise statement was finally issued.

This is what I call "institutional scientific misconduct." We also ran into it with the Nuclear Regulatory Commission, the Department of Energy, the Environmental Protection Agency and the National Council on Radiation Protection and Measurements, who refused to properly consider the relevant data and winked at badly flawed reports that appeared to support their position. Cutting off research they didn't like and selecting like-minded individuals for their committees, they managed to sustain the status quo.

Scientific misconduct by some individual scientists was also discernible. Simple refusal to acknowledge beneficial effects was the most common. They wrote, for example, "(since) there is no reason to expect radiation to *decrease* cancer mortality" and went on to ignore such data or even to count all health effects as detrimental. They worked mostly at higher radiation levels and then stated that any effect or absence of effect at lower levels would be too small to observe, which is true only if you assume the LNT applies—a classical case of "begging the question."

The word got around that if researchers wanted funding, they should be counting dead mice, not looking for happy mice.

The prevalence of such conduct has become sufficiently blatant and widespread in the field that some formal scientific misconduct charges may have to be filed. These have been effective in other areas where a political agenda tends to erode scientific integrity.

Energy Secretary Says: "We're Killing People"

When the Nuclear Regulatory Commission was formed in 1975, it was given full responsibility for regulating nuclear facilities. This left the new Department of Energy (DOE) free to promote and encourage the development of nuclear technology without any regulatory conflict. But that did not prevent DOE from stretching the LNT premise to state in a news release on June 3, 1997, that "after six years of study and analysis" it concluded that 23 people will be irradiated to death as a result of the trivial radiation doses from shielded shipping casks carrying radioactive waste across the country. There is no scientific basis for such a statement, and it raised baseless fear of fuel shipments.

Then, on July 15,1999, an official DOE news release had this ominous paragraph:

> **Radiation-Induced Cancers**. We estimate that over the next 30 years, there will be between 250 and 700 radiation-induced cancers among DOE contractor employees, of which about 60 percent will result in death...

But the worst was yet to come.

On January 29, 2000, Energy Secretary Bill Richardson released a bombshell. As reported in the *Wall Street Journal*, "After decades of denials, the government is conceding that since the dawn of the atomic age, workers making nuclear weapons have been exposed to radiation and chemicals that have produced cancer and early death." Richardson was quoted as saying, "This is the first time that the government is acknowledging that people got cancer from radiation exposure in the plants... Justice has finally come; the government is for a change on their side and not against them." The article ended by saying that these conclusions were all based on previous publicly available reports. "None of the research was done specifically for this study."

Since this conclusion contradicted every valid scientific study I'd seen, I called the Secretary's office and asked for a copy of the report on which the statement was based. I was told that the report was not available. So a formal request for the data was written by RSH on February 1, and the Assistant Secretary for Environment, Safety, and Health, David Michaels, wrote me to say that the report was prepared by the National Economic Council (that's right; by economists!) based on "studies previously dismissed." When the economists have concluded their efforts, "we anticipate a public robust discussion involving not only the issues raised in the report but also whatever recommendations come out regarding compensation for DOE workers."

No such public discussion ever took place. The industry and the scientific community were strangely silent. The Secretary told reporters that the responsible officials knew that workers were being killed, but they lied about it and covered it up for fifty years. Still no reporter queried the Secretary. Congress quickly passed a "Sense of Congress" statement declaring, "Since World War II federal nuclear activities have been explicitly recognized by the U.S. Government as an ultra-hazardous activity." No mention was made of the fact that all valid studies show that nuclear workers have *better* health, *less* cancer and *greater* longevity than other workers.

Most egregious was the unprovable and indefensible statement, "Furthermore, studies indicate that 98% of radiation-induced cancers within the DOE complex occur at dose level below existing maximum safe thresholds." Reading the transcript of the Congressional hearings, I saw not a single voice of doubt or question as the measure passed without objection. No one wanted to be seen as opposing handing out money to the "cold war heroes." Since no one can prove that an illness was caused by low-dose radiation—none has ever been detected—the decision as to compensation in each case will apparently be based on whether the illness *could have been caused* by radiation. The list of illnesses and symptoms that might possibly have such an origin grows almost daily, so the number of eligible beneficiaries increases correspondingly.

Shortly after the election of a new administration, Secretary Richardson accepted a position as Director of the anti-nuclear political action group Natural Resources Defense Council, who announced that they considered that he had been a good Energy Secretary.

We in RSH had had little response to our efforts to get working scientists to speak out on the discrepancy between data and policy. They saw our efforts as endangering their funding. "That's not how I get tenure for my people," was the response from an official of the National Cancer Institute when we asked for help in getting low-dose irradiation treatment for a patient with an "untreatable" blood cancer.

We appealed formally to appropriate organizations to investigate this gap between the policy and the science. We wrote detailed documented letters to the U.S. Department of Energy, the U.S. Nuclear Regulatory Commission, the U.S. Environmental Protection Agency, the advisory commissions and panels, and the congressional leaders who had expressed interest in nuclear technology.

None of this led to effective action until Senator Domenici asked the General Accounting Office to investigate it. For a year we worked hopefully with the director of that investigation, but the ensuing report was toothless. Virtually all the material we had provided was ignored. Senator Domenici arranged for the Department of Energy to undertake research to resolve the matter, but this money was diverted to a multimillion-dollar, 10-year program on genome functions and cell cultures, designed to yield no animal or human information in the foreseeable future. It was clear that more decisive action was needed.

Suing the U.S. Environmental Protection Agency (EPA)

About that time (March 2000), I received a call from Alan Pemberton, a lawyer from the prestigious firm of Covington & Burling, asking if he could use a paper I had written on the LNT issue. I discussed the problems we were having in reaching any resolution to that situation, and asked if his firm ever did any *pro bono* work (i.e. for free). He said,

"Of course. We do quite a bit." I asked who in the firm made the decision to take a case on that basis, and he said, "I'm the chair of the pro bono committee."

"Have I got a deal for you!" I replied, and we talked about the possibility of approaching the LNT problem from a legal, rather than a purely scientific, standpoint. He responded positively and said he would check whether his firm could help us. In a week or so he returned with a big yes and turned me over to Kipp Coddington, one of their lawyers who also had an engineering background and was quite familiar with the radiation question and with EPA.

RSH's many discussions with Coddington were fruitful. He advised that we find a particular legal action that could be challenged, rather than seeking to challenge the overall philosophy. This made sense to us, and we found that the EPA had just proposed a rule on the permissible levels of radioactivity in "Primary Drinking Water." The proposed rule had a deadline for public comment of June 20, 2000, and by working diligently we were able to complete our comment, with a nearly foot-thick stack of legal and technical attachments, in time for me to load twenty copies of it into my van and carry them over to the EPA office an hour before the close of business on the last day.

The next step was to wait for EPA's response. EPA was under court order (unrelated to our action) to issue a final rule by the following November, so we knew that EPA could not delay its response for years, which often happens. The response, as expected, did not address our basic objection, namely that in using the LNT premise to establish permissible limits, EPA was "arbitrary and capricious" in ignoring the best peer-reviewed science that the law required it to use. The EPA rule set goals for each nuclide at zero and then required that operating levels be reduced as near zero as technologically feasible—a continually elusive target. So the next step was to petition the federal appeals court to review the rule (and hopefully send it back for revision).

On January 4, 2001, just 15 days before we were to file our petition with the court, I got an email from Coddington. I had over 40 previous emails and numerous phone calls from him on this case, but this one was different. Its message was simple:

> We cannot represent you in this matter for a variety of reasons that are too difficult to explain here but are unrelated to the merits of your case.

I was never able to determine why his firm had pulled out or who put the pressure on them. Neither Coddington nor Pemberton would answer my questions, and I was left to scramble for a lawyer to pick up the pieces. Luckily, John Ferguson, who worked with my engineering firm, MPR Associates, agreed to assign one of his lawyers, Mike Wigmore, to help us, and he took over filing the petition. I had previously met with lawyers from the American Water Works Association, the National Mining Association and the City of Waukesha (Wisconsin) Water Utility, each of whom had submitted sharply critical comments to EPA. Mining and water treatment operators saw that the rule would suddenly convert each of their small, local operations into federally controlled radiation handling facilities, with implications they could only begin to see. The mining association joined us in the suit, as did Waukesha, but the Water Works Association held back. Waukesha persuaded six other local water-treatment facilities to join in. And the Nuclear Energy Institute (NEI) also filed as petitioners at the last moment.

This last one surprised me, because I had spent two years trying to get NEI to join with RSH to challenge the LNT. Joe Colvin, NEI President, told me that he and his members did not think this was an issue that could be settled in the foreseeable future, and they were staying out of it. We had also approached the Joint Defense Group (JDG), the utility lawyers who work together to defend utilities being sued for radiation-induced injuries. They told us the insurance companies used to just settle these cases out of court without regard to the merits. So the utilities set up the JDG to fight them.

Our suggestion to get at the root of the problem, to challenge the indefensible premise that low-dose radiation is harmful, met with indifference. They apparently see no reason to contest the present situation with its steady flow of cases.

What would it mean to win such a lawsuit? Would a judge settle a scientific question in court? I've been told that judges will not choose between opposing scientific viewpoints and say, "That one is better." But what I have seen judges do is to determine that an agency did not follow proper procedures in arriving at its scientific decisions and therefore, acted arbitrarily and capriciously. The matter would then be returned to the agency to be dealt with properly. For example, the EPA used the LNT premise to set a limit for chloroform in water. A judge ruled that no scientific basis had been cited for believing that chloroform acted that way. He struck down the ruling and remanded the matter to the agency for revision. In another case, involving the carcinogenicity of second-hand tobacco smoke, the Conclusion of the 90+ page five-year case stated:

> "EPA publicly committed to a conclusion before research had begun; excluded industry by violating the Act's procedural requirements; adjusted established procedure and scientific norms to validate the Agency's public conclusion; and aggressively utilized the Act's authority … to establish a de facto regulatory scheme."

The Court also noted that, in conducting its risk assessment:

> "EPA disregarded information and made findings on selective information; did not disseminate significant epidemiological information; deviated from its Risk Assessment Guidelines; failed to disclose important findings and reasoning; and left significant questions without answers."

On the basis of such history, we had hoped that the legal system might provide us what we were unable to get from the scientific community or the executive branch of the government: an unambiguous and unavoidable requirement to competently and objectively answer the question: Is low-dose radiation harmful? In this endeavor, we could silently pray with the bailiff: "God save the United States of America and this honorable court."

On February 25, 2003, the Court handed down its decision: "We conclude that … EPA complied with the requirements of the SDWA and the APA" (the applicable laws). Some of the petitioners are considering appealing to the U.S. Supreme Court. One told me he had already spent a million dollars on this case and appeal would probably cost another million. "But it will cost me $70 million to comply and make a nightmare out of running a little sewage plant," he complained. "And the idea that no one should challenge a federal agency is one this Supreme Court might want to jump on." Meanwhile, RSH is exploring amending the law, in which there was already some interest in Congress. I hope the next edition of this book will be able to report a happy ending to this saga.

Why Is This Important?

I found that most of the nuclear movers and shakers did not place the problem of low-dose radiation high on their priority list. They viewed our efforts to reform the situation as mildly commendable, like sending money to the Red Cross. But they considered the whole subject rather arcane and theoretical compared with other more urgent items confronting them. Why then do I consider it so important? What difference would winning this case make in the real world?

Nuclear operations, and specifically nuclear radiation, are widely viewed as presenting an unprecedented hazard to the human race—one we should not accept if there is any possible alternative. This is expressed many ways, but the underlying factor in each case is the argument that no amount of radiation is harmless. Thus, if there was any exposure to radiation, a worker who gets cancer concludes that radiation must be the cause. The law often agrees with him, and nuclear employers often pay off without questioning it.

LNT advocates who call themselves pro-nuclear claim there is no practical difference between saying that low-dose radiation poses no risk, and saying that the risk is less than other risks routinely accepted and therefore, should be tolerable. But this is, in fact, a black-and-white difference. Many people argue, and some courts agree with them, "it is not for *you* to say that *I* must accept an increased risk of cancer just because you find it tolerable. I say you have put me at risk without my consent, and I want compensation."

We don't apply the LNT philosophy to any other hazards. We don't ascribe deaths to the highly-toxic trace elements in our vitamin pills, such as selenium. In fact, we pay for them believing they are beneficial. The same reasoning should hold for radiation. We pay a high price for treating radiation differently.

Thousands of tons of nickel and other valuable metals used in the nuclear industry cannot be recycled because of trivial contamination. Families living near the burned-out Chernobyl nuclear plant are showing record levels of alcoholism, clinical depression and suicide because they can't return to their ancestral homes. They are told the land is "contaminated," although the radioactivity is generally lower than first-class real estate in Denver.

Japan has been forced to cut back and rethink its national commitment to nuclear power, because it recently experienced the worst nuclear accident in its history. This accident killed two factory workers at the uranium enrichment facility and "exposed over 600 members of the public to radiation." No one notes in repeated use of this phrase that the public radiation exposure was less than those individuals might encounter from variations in natural background and was too low to cause any detectable harm.

Some hospitals have had to close down their life-saving nuclear medicine facilities because of the burdensome regulations and uncertainty about accountability and disposal. Lives are lost because people have been warned away from treatments such as mammograms that involve radiation. Firefighters tell us that lives are lost because people won't buy smoke detectors with radioactive sources. Tens of thousands are said to die from respiratory illnesses caused by fossil-fueled power plants. Plans for fighting global warming and other ecological damage are distorted to exclude nuclear power, the most benign power source. "The waste problem" turns out in the end to be concern for reducing still further radiation levels that are already harmless.

Once the idea is established that any potential source of radiation must be treated as an extraordinary hazard, then spent fuel and other radioactive material become objects of terror—"mobile Chernobyls" in the words of the media. Although such shielded containers simply cannot hurt anyone, we are told that we must protect them from terrorist attacks and track them with extreme measures as they are trucked across the country. We are urged to ring nuclear facilities with extra guards and anti-aircraft batteries, even though it would be impossible to cause a release of radioactivity through any credible attack scenario on the reactor structure or the fuel storage pools that would create a serious public health hazard. This conclusion is described in a heavily documented, peer-reviewed pair of papers in *Science*, (September 20, 2002; January 10, 2003) that I co-authored with eighteen other Engineering Academicians. More life-threatening as well as more likely would be an attack on the chlorine storage tanks at a local water-works, an oil refinery, a natural gas pipeline, or even a neighborhood automobile filling station.

Even natural radiation that we have lived with healthily for countless generation is now characterized as hazardous, requiring government regulation and control.

While agreeing that low-level radiation is "probably harmless," critics argue that "ALARA (reducing radiation 'as low as reasonably achievable') will always be good public policy." But of course if we believe the scientific data, we must conclude that such policy is not only expensive, burdensome and unnecessary, it is counter-productive. It causes harm, not benefit. Until we understand that very important basic fact, we will always be fighting the radiation bogeyman to the detriment of the public welfare.

8. Going Civilian

Declassification and Public Review of Safety

Background

The first few years of the 1950s were frantically busy and very confusing for Captain Rickover and his Naval Reactors people as one challenge after another piled up. President Truman authorized the *Nautilus* on August 8, 1950, and one week later the Chief of Naval Operations asked us to explore the feasibility of developing a nuclear-powered aircraft carrier, a propulsion plant requiring twenty times the power output of the submarine. That same month we broke ground for the full-sized submarine prototype plant at Idaho, but it was a full year later before we were finally able to sign contracts with Westinghouse for the power plant and Electric Boat Company for the ship and the prototype hull. Two months after that in October 1951, we were asked to study design criteria for a high-speed submarine.

On May 31, 1953, Captain Rickover arranged for Atomic Energy Commissioner Thomas E. Murray to open the main turbine throttle of the *Nautilus* land prototype plant, and they both solemnly walked to the back of the hull to watch the propeller shaft turn over for the first time. Murray, the only Atomic Energy Commissioner who was an engineer, had been especially supportive of Rickover's efforts to achieve this goal, and Rickover knew he would be particularly appreciative of its significance. Years later, Rickover remarked that he had not known ecstasy many times in his life, but the two occasions he recalled were, first, his appointment to the Naval Academy; and second, watching the propeller shaft of the first atomic engine turning over and knowing that his dream of nuclear power was within reach. We were all relieved to have the first solid evidence that this unprecedented machinery might in fact work as intended, and on June 25, the plant reached full power for the first time. Rickover insisted that the plant remain at full power for 100 hours, and posted a chart of the Atlantic Ocean in the engine-room, on which the crew plotted, hour by hour, the course of a simulated nonstop transatlantic submarine voyage. Rickover had been passed over for promotion the previous year and told, at age 52, that he must retire from active duty by the following June. But after a great public protest and pressure from Congress, Rickover was promoted to rear admiral on July 1, 1953. (I describe the 100-hour run and the promotion struggle in more detail in *The Rickover Effect.*)

Before the prototype plant had proved it could produce power, President Eisenhower had already authorized construction of the world's first commercial atomic power station. Eisenhower wanted this to be a fully commercial program with no military flavor whatsoever, but it was clear that no one else was in a position to undertake this task. Shortly thereafter, on July 9, with the strong support of Commissioner Murray, the newly promoted Admiral Rickover was assigned the job of converting nuclear power to an unclassified, civilian, commercial status.

But we were living in a militarized world. By the end of July, the Korean War was winding down, and two weeks later the U.S.S.R. astonished the world by testing its first hydrogen bomb, a thousand times more powerful than the weapon that had incinerated Hiroshima. Over the next few years, the Soviet turned down Eisenhower's Open Skies proposal under which all nations could assure themselves by direct observation that no nation was carrying out a clandestine nuclear arms program. Instead, Russia launched intercontinental ballistic missiles and the world's first artificial satellite, and began an unprecedented buildup of nuclear-powered submarines. The arms race was at full pitch, and our Naval Reactors program was primarily a military one. We were constantly reminded that the technology behind these submarines was a precious military asset and must be rigorously guarded by secrecy. Since the security measures set up by the Manhattan Project were still largely in force, it had not been difficult for us to operate as a highly-classified military program.

Now we were to create a commercial atomic power plant, operated by a private utility, selling dependable electricity to its customers—a plant that became known as Shippingport, after the little Pennsylvania town where it was built, 25 miles down the Ohio River from Pittsburgh.

Suddenly Civilian

The idea of developing an industrial nuclear power technology did not start with the Shippingport project. From the beginning, Rickover had envisioned his naval reactors program as an engineering project, to be built by industrial firms, not a science project to be built in a laboratory. Moreover, he never considered his objective to be limited to the creation of a single experimental test vehicle, like the Wright brothers' first airplane. He foresaw a global nuclear-powered fleet that would require suppliers and support facilities on a worldwide basis. That thinking permeated everything we did.

President Eisenhower had similar aspirations for his civilian nuclear power program. He did not want a one-shot demonstration. He wanted us to disseminate widely, on an unclassified basis, all the technology and design information necessary to enable industrial firms to bid competitively and intelligently on the thousands of pieces of the package that would ultimately lead to a working commercial electric power generating station. This would also equip others to follow in our footsteps. There was considerable concern outside the program that Rickover would not take that part of his assignment seriously, that he would merely stick a civilian nameplate on the aircraft carrier plant and substitute an electric generator for the ship's propeller. They didn't know Rickover. He did not do anything half way.

Starting in 1946 at Oak Ridge, long before Shippingport, Rickover's people had put out a series of technical handbooks, spelling out in detail the scientific and engineering fundamentals of nuclear technology. The process of organizing, writing and documenting a new area of research and technology, and submitting it to peer review, can be as much a creative process as the original work, and it results in more than a mere snapshot of something already in existence. Some of the Naval

Reactors handbooks have been used for over 40 years as textbooks, training manuals or source books in civilian installations all over the world, and are still being cited in the scientific literature. All this was already in full swing even before Rickover left Oak Ridge in 1947, and continued unabated as we developed new information in the years that followed.

But that was just the beginning. For example, many of the industrial codes and standards now used as basis for the civilian nuclear power industry were first developed for naval reactors. The Boiler and Pressure Vessel Codes, several important welding standards, and a large number of the standards issued by the American National Standards Institute (ANSI) can trace their ancestry directly to that early military work.

Again, this was no trivial task, and its significance was immense. It meant that Rickover and his people were setting the standards for a wide variety of industrial products and practices, not only for the nuclear power industry to come, but for much of the rest of the world's heavy machinery components and systems as well. Serving on a national standards committee requires endless hours of meetings, reading, writing, and correspondence. It is one of the important ways that an engineer does *pro bono* work, uncompensated and largely unsung, for the public good.

Another technological accomplishment of the naval program was the establishment of a nationwide system of nuclear training programs, schools, curricula, texts and instructors, covering basics such as math, nuclear physics, radiochemistry, heat transfer and fluid flow, materials and metallurgy, and more specific topics such as reactor plant dynamics and control, health physics and prototype training and qualification. These training materials became the basis for the civilian programs that followed, and over 100,000 nuclear-trained naval officers and enlisted personnel provided a priceless source of experienced personnel for the new civilian industry. In addition, we already had Westinghouse and General Electric running AEC laboratories devoted solely to our program and subcontracting mostly to industrial firms, rather than universities and government laboratories.

Creating Zirconium and Hafnium Industries

The first big challenge for building any of our reactors was to create an open, competitive zirconium industry to provide material for the reactor core structure. This effort was already under way for the *Nautilus* program, but the civilian project gave it added scope. Although zirconium ore is widely and abundantly available, only a shoebox-full of zirconium metal existed in the whole world, and it was too brittle to fabricate and too impure to use in a reactor.

For an agonizing year and a half, Rickover pleaded with the Atomic Energy Commission to let him take over zirconium procurement, which was bogged down in the AEC's Production Division. AEC management agreed that a zirconium production capability was needed, but Rickover's was only one of many reactor programs, and they didn't see why his program should be handled differently. (This may be one of the reasons that none of the other programs ever amounted to much.) In August 1950, Rickover was finally given responsibility for zirconium procurement. At that point, he had less than four and a half years to get the *Nautilus* to sea. Before that happened, he would be asked to undertake the civilian reactor project.

Rickover got together with his people to size up the zirconium problem. "What do we have to do?" he asked, with his usual directness. He always asked the most basic question, as if he had never heard the word zirconium. Most of us (including me) usually want to show we know a little bit about the subject, so we take the first answers for granted. Sometimes this leads to real trouble, when you later find that the answers to some of the first questions (the ones you didn't ask) turned out to be different from what everyone had been supposing. So Rickover always started at the very beginning. He met with Lawton Geiger, his manager of the AEC Office at the Bettis Lab, Jack Kyger, his Technical Director at that time, Harry Mandil, his senior reactor plant and mechanical engineer, and Bill Wilson, whom Kyger had brought with him from Oak Ridge.

"The problem is not just one of scaling up an existing process; there *is* no existing process," explained Kyger.

"First, we have to learn how to make a reasonably pure metal in sufficient quantities to test its properties."

"What properties? What tests?" asked Rickover.

"Corrosion tests," said Mandil. "And mechanical properties; strength, flexibility—that stuff."

"And nuclear," added Kyger. "We don't know much about its neutron absorption and scattering. When you finish with the mechanical and corrosion tests, you'll probably find you can't use the pure stuff; you'll have to make up some kind of alloy. So then you have to go back and test the alloy."

"And you gotta go back and forth," said Wilson. "You finally find an alloy that's strong enough, but it's too brittle or it corrodes, and you try again."

"What quantities are we talking about? How much will we need?" asked Rickover.

Geiger jumped in on that one. "The AEC has been complaining that the procurement program needs a firm commitment, a requirement of a specific amount for a specific purpose. I've been working with Kyger and with the Bettis people on that, and we believe that we need at least thirty thousand pounds, just for the prototype plant. The ship's requirement will be on top of that."

By March 1951, they got the AEC to issue a formal requirement for zirconium and sent it to twenty-six potential suppliers. By that time it was clear that hafnium, an element that usually occurs as an impurity in zirconium, would also be needed to act as a neutron absorber for the reactor control rods, so it was included as an additional requirement. Thirty-four companies attended the contract conference, complained about the tight schedule and objected to the fixed price arrangement. So a revised invitation was issued to thirty-five firms, but no responsive bids were received.

This situation provided a legal basis for negotiating the best possible deal with a single company, which resulted in a contract with the Carborundum Metals Company for two hundred thousand pounds per year for five years—a nice, even million pounds. Other companies and the federal Bureau of Mines were involved in other aspects of the purification and production process, and work that had previously been done at AEC laboratories was moved into commercial firms. Eventually other production companies were brought into the business, leading to a highly competitive industry producing tens of millions of pounds per year of zirconium and hundreds of thousands of pounds of hafnium. Zirconium became the metal of choice for nearly all of the world's nuclear power plants at about $5 per pound.

Dan Kimball, the Navy Secretary who pinned a medal on Rickover for the *Nautilus,* testified before Congress in 1952. He told them he asked Westinghouse how they had been able to produce zirconium so quickly and so well, when other companies had all failed at it. "Rickover made us do it," was the reply.

Procurement of Equipment

As the program grew, the procurement of equipment from competitive commercial sources became a major operation. To avoid interference with the research and development operations, Rickover had Westinghouse set up in 1956, at Cheswick, Pennsylvania, a separate department solely for procurement, away from the Bettis Laboratory site. This was not a standard purchasing department, but was a highly technical operation with senior technical people charged with educating industry as to what was required to make suitable nuclear equipment and developing the necessary standards and specifications. By the end of 1958, the new department was working with 400 suppliers, 55 of whom had contracts of $100,000 or more, and 21 had contracts ranging from $1 million to $15 million. To put these numbers into perspective, note that the entire research, development and construction cost of the fullscale *Nautilus* prototype power plant was only $178 million. In 1959, Rickover had General Electric set up a similar facility for the *Seawolf* submarine project, already underway.

Going Public

Despite our experience in working with industry in the naval program, the transformation from a military mode of operation to a civilian one was a big adjustment. Like the island of Madagascar or Australia, the realm of military research, development and procurement has developed in isolation from the rest of the world. It has, therefore, developed some unique flora and fauna not seen anywhere else. It has its own versions of the kangaroo and the duck-billed platypus, not to mention scores of lemurs of size, shape and color not even imagined elsewhere. After living many generations in this surreal jungle, its denizens come to follow procedures whose special characteristics have to be experienced to be believed, and they speak in strange acronyms and exotic jargon.

The Navy does this best. Who can match such melodious phrases as "us nussle" for the United States Navy Underwater Sound Laboratory or "us nav raddle de flab" for the U.S. Naval Radiological Defense Laboratory? When a naval officer was not at sea, he was "on the beach," even if he worked in a large office building in Chicago. If you went into that building and asked how to get to his office, you may have been told "On the second deck, sir. Right up the marble ladder there". And if you're looking for a rest room, be sure to ask for "the head." I remember a man who worked on "JAN Specs," combining the various supply catalogs of the Army and the Navy into joint specifications. (There was no Air Force in those days.) One day he announced in triumphant tones that he had discovered that the Army listed an item as "pong balls comma ping" whereas the Navy listed them as "balls comma ping pong," thus frustrating his attempts to merge them. Having found the problem, he could now correct it, mark up another victory, and move on to the next target.

So, although we had already begun to develop many of the same kinds of physical and organizational facilities needed to do the civilian work, there were basic differences that had to be overcome. There was a massive declassification program necessary, to pull the pertinent technical and organizational information out of the SECRET files and put them into a form suitable for wide public distribution, and then process them for declassification. The next difference that struck us was the continuous public accountability of the civilian operation. In our military work, we were always strictly accountable to the Navy, the Atomic Energy Commission, and the Joint Committee on Atomic Energy and other committees of Congress such as Military Affairs and Appropriations. We were scrupulous in meeting these responsibilities, but it was much more of a one-on-one or personal affair than would be feasible in the civilian world. In the military projects, we would get a call from one of the many agencies involved, asking a question or imposing a requirement, and we would go over and talk to the person directly. Sometimes we were required to do something we thought was unnecessary, but generally it was just a matter of clarifying what we were up to, and no change was necessary. We were nearly always dealing directly with the person who had the inquiry, and we could generally get it quickly clarified and resolved, one way or another.

Now, in the civilian world, things were different. At first, the differences were minimal. There were no other significant reactors and few nuclear bureaucrats; the few low-power research reactors that did exist or were under construction were being done at the AEC's national laboratories under the AEC aegis. It should be noted that from the beginning, the AEC's own facilities never underwent the same kind of public review process that private reactors were subjected to, but Rickover determined to go the whole way with Shippingport—that is, to do it under the full glare of public scrutiny. In view of the nature and purpose of our project, it would not have made sense to do it any other way, even if we had wanted to do so.

We were really paving the way for the commercial safety review process to come. In addition to the Naval Reactors Branch, the AEC's Division of Reactor Development had set up Army and Air Force Reactors Branches, and in due course they set up a Civilian Reactors Branch. The Civilian Branch, under a lawyer named Harold Price, was responsible for developing procedures and criteria to be used for reviewing and approving the operation of the civilian reactors that were expected to come in the wake of Rickover's Shippingport plant. In December 1974, in order to separate its regulatory from its promotional functions, Congress split up the AEC into the Nuclear Regulatory Commission (NRC) and the Energy Research and Development Administration (ERDA). ERDA was later renamed the Department of Energy (DOE). The NRC then took on all responsibilities and functions associated with regulating, evaluating and approving all nuclear reactors except those being built by and for the Government leaving DOE free to promote nuclear energy. (It's ironic that under President Clinton, DOE, the *promotional* agency, became almost pathologically afraid of being accused of promoting nuclear energy. Unlike the Federal Aviation Administration or the departments of Labor, Agriculture or Commerce, the DOE acted as if any evidence of advocacy for its clients would be unseemly.)

Figure 8.1 Air view of Shippingport, the world's first commercial atomic power plant (*Westinghouse*)

At first, Hal Price and his people talked with us to determine how we handled review and approval of the naval plants, what kinds of questions we used in examining the operators, and how we envisioned the Shippingport plant would be evaluated. But in time his organization grew and he developed procedures and criteria of his own. He shared his thinking with us as it developed, and we were glad to show him how we operated. By the time Shippingport was ready for review, the review and evaluation process, in concept, was approaching the system that exists today (see Figure 8.2). The fact that Price was a lawyer, rather than a scientist or an engineer, foreshadowed the kind of approach that would be taken in evaluating and controlling nuclear plant safety.

Reviewing Civilian Reactor Plants for Safety

These days, the first step in getting approval to build a nuclear power plant involves submitting plans to the U.S. Nuclear Regulatory Commission (NRC), along with a Safety Analysis Report and an Environmental Impact Statement. Each of these reports consists of twenty to thirty volumes of detailed data and analysis, which the utility company and its contractors have prepared for this purpose at a cost of several million dollars.

The NRC is a public agency that has no responsibility for seeing that a plant is built on time or within budget, or whether it is even built at all. The NRC is intended solely to be the people's watchdog on all nuclear activities, responsible only for public safety, with no conflicting obligations.

For a period ranging from a few months to several years, the NRC negotiates with the utility company, typically requiring further calculations, and often compelling certain design changes, until it is satisfied that the plant can meet all applicable safety requirements. In most cases, these requirements have been known all along to the plant's designers, and it has been their intent to meet them, since the plant cannot be approved if it falls short. But it takes this prolonged period of review and modification for the regulators to convince themselves in detail that the plant design satisfies the requirements, and the plant designers to agree that the modifications negotiated do not introduce other unacceptable problems. This safety review covers not only the technical features of the plant, but also factors as the background and capabilities of the utility's management personnel, the company's fiscal soundness and freedom from foreign control, and estimated impact of traffic, noise, and materials handling on surrounding people, vegetation, and wildlife.

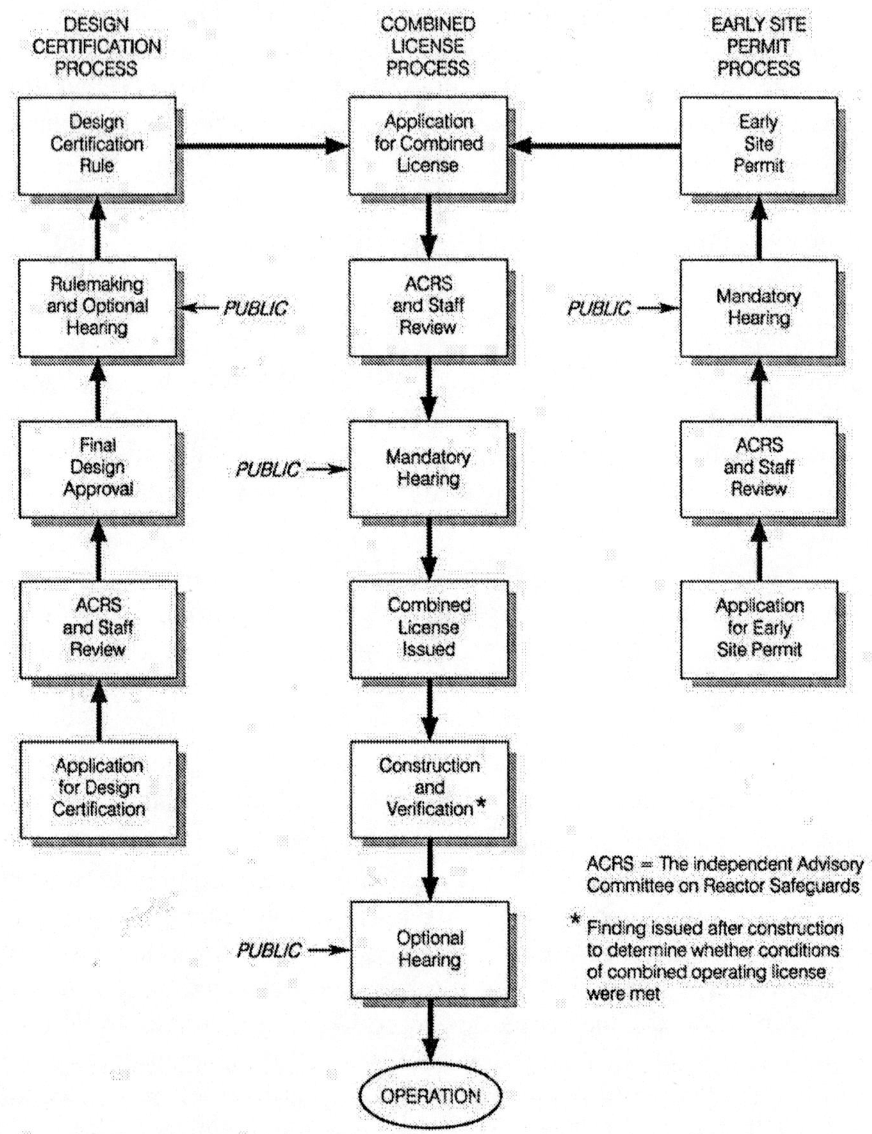

Figure 8.2 Flowchart of the reactor licensing process (*DOE*)

With the National Environmental Protection Act of 1969 (NEPA) and the establishment of the Environmental Protection Agency (EPA), the scope of this review became even broader, covering such things as aesthetics and other hard-to-define aspects of the quality of life.

Public Participation in Nuclear Power

When finally satisfied, the NRC calls for a public hearing. These hearings are run by an independent Atomic Safety and Licensing Board, an *ad hoc* panel of judges and technical personnel acting like a combined judge and jury. This public hearing process permits anyone who has questions the NRC may not have raised to bring them up publicly for resolution. Unfortunately, the image created by this process is that the NRC is in bed with the utility and is acting almost as its defense attorney. This belies the months or years of aggressive adversarial examination that preceded the public hearing.

All correspondence, reports and testimony are available in public document rooms the NRC operates all over the country, including one near the proposed site of the plant under scrutiny. All manner of documents are available there, along with clerks to help locate documents, and copying machines for public use. In addition to the mainstream reports, minutes and correspondence, every crackpot letter, anonymous allegation and internal memo that was considered, evaluated, and turned down, and all criticisms and complaints are available for viewing and public re-evaluation.

One of the first nuclear plants to which the NEPA rules were applied was the Baltimore Gas & Electric plant at Calvert Cliffs, Maryland. I sat in on that hearing in 1970, and it was a fascinating experience. It had a strong flavor of a Norman Rockwell painting or a William Saroyan play, about a good old-fashioned American town meeting, with friends and neighbors having their simple, blunt say. Some were eloquent, some bumbling; some were insightful, some were vague. I remember a real estate developer who seemed to have stepped right out of "Guys and Dolls," apparently just an interested spectator to the proceedings, who suddenly got up and delivered a wonderful light-hearted diatribe against environmentalists. "I went tru all dis in Flaridah in da fordies," he announced. "If doze guys had had dere way, ya wouldn'a had Miami. You'da had nuthin' but ducks and mosquitoes. Ya ever try ta raise any tax money offn a duck?"

He was allowed to finish his statement, he sat down satisfied that he had been heard, and the next scheduled speaker was called up. The public was encouraged to ask questions, and even short speeches like the realtor's were permitted. It was all very open and pleasant, and nobody was steamrollered.

Protesters sometimes complain: "But they always approve the plants in the end. So the whole process is a sham!" The answer to that complaint is straightforward. The NRC is convinced that it is possible to build and operate a nuclear plant safely, and they have laid out the criteria such plants must meet to be considered safe. Therefore, the review process takes the form of insuring that the design of each particular plant does indeed meet these requirements; and wherever it appears to be deficient, the NRC will insist that appropriate changes be made before it is willing to accompany the utility to the public hearing.

With few exceptions, all American plants are pretty much like those already reviewed and approved. The license review procedure is an "open book exam" in that the whole review process is open to the public, and the record is available to other utilities whose plants will be reviewed next. So it's not surprising that they eventually "pass the exam." To me, the surprise is that it usually takes several years to do so, as interveners raise again questions that have been considered and resolved many times before.

Living with Regulation

The mad scramble to build nuclear power plants reversed itself about 1970, and the last American plant was ordered in 1974. Costs escalated to undreamed-of heights as protracted public hearings dragged out construction times, and interest rates above 20 percent made delay cost up to a million dollars a day. An entire industry of interveners had grown up, making careers of trying to delay or completely block the construction of one nuclear plant after another. The industry itself often bumbled, delayed, gave incorrect answers, or tried to cover up an embarrassing situation, and the professional anti-nuclear lawyers made the most of it. The almost-religious fervor driving some of the anti-nuclear interveners was akin to that exhibited by abortion clinic protestors. In addition, the interveners were often abetted by local Public Utility Commissioners who were ideologically opposed to nuclear power. After the Three Mile Island accident in 1979, the whole industry looked moribund.

This atmosphere made even straightforward technical and management discussions difficult. Employees and consultants had constantly to keep in mind that their letters or reports would be in the hands of high-priced lawyers for anti-nuclear activists who were suing to block construction of the plant, and this could happen before the addressee had even read them. This made it difficult for a consultant or a utility employee to report bluntly to management that a design or fabrication detail was deficient. I know of no other large enterprise that had to proceed so long under such intense, continuous, hostile public scrutiny. In this situation, the procedural jungle proliferated like the rain forest in a good season, peaking shortly after the TMI accident, as the lawyers and the scientific theoreticians sought to ensure safety with an ever-increasing deluge of calculations and edicts.

Nuclear plant construction coasted to a standstill, and in the silence that followed, some introspection inspired by new management and congressional inquiry ensued. NRC procedures were streamlined and were redirected away from treating all theoretical scenarios equally and focused more on issues with significant safety implications. But this laudatory situation has yet to be tested on a new plant, since no new plants have been started in nearly thirty years. The experience in reviewing license extension applications has been encouraging, however, and several plants have now added twenty more years to their operating licenses.

Figure 8.3 shows the growth in the number of applicable federal regulatory documents and professional standards, and the increase in plant cost during the same period. Figure 8.4 shows the growth in required documentation, as measured by the number of pages of correspondence and reports actually created in connection with plants in the 1960s vs. the 1970s.

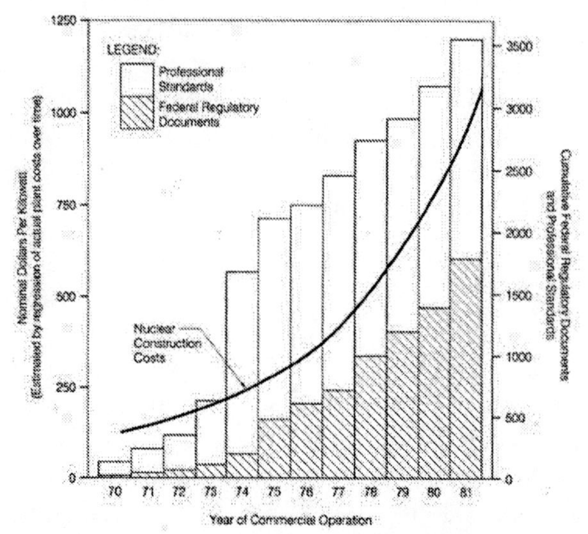

Figure 8.3 Growth in cost of nuclear plants and growth in number of regulatory documents (*Westinghouse*)

Item	1960s	1970s	% Increase
Pages of correspondence			
- Transmitted	12,400	55,000	340
- Received	8,600	17,000	98
Safety Analysis Report (Volumes)	2	13	550
Environmental Reports (Volumes)	2	22	1,000
Public Hearings (Days)	2	16	700
Number of Quality Manuals	1	6	500
Pages of Calculations	1,400	70,000	4,900
Number of Drawings	2,200	45,000	1,945
Number of Specifications	230	490	113
Number of Vendor Drawings	37,000	90,000	143

Source: Licensing, Design & Construction Problems: Priorities for Solution, January 1978, Exhibit 9 (Atomic Industrial Forum)

Figure 8.4 Growth in required nuclear plant documentation (*Author*)

In November 1981, Marcus A. Rowden, a lawyer long familiar with nuclear energy regulation, gave a talk deploring the situation and stating that, although the law in this case created some problems, much of the mess was created by the NRC and could be changed by the agency itself:

> It is a modern-day paradox—and a costly one—that a technology from which we expect innovation and demand the highest standards in its commercial use has imposed upon it a licensing process which is so archaic and counterproductive ... the trial-by-combat nature of the hearing makes it an inhospitable forum for arriving at decisions resting on technical discourse and judgment....Dissatisfaction with the current hearing process is one of the few things which all licensing protagonists—applicants, interveners, the NRC itself—share in common...

He goes on to recommend, "The role of the [NRC] staff as a party in licensing hearings should be eliminated." As he pointed out:

> ... the staff's most publicly visible function is as a seeming "sponsor" of [the applicant]. This is not merely a cosmetic deficiency; it is misleading, since the preceding 2–3 years of application review have been ones of arms-length (and often adversarial) relationship.

He cites U.S. Senate report No. 113, 97th Congress (1981) that estimates of the cost of this licensing process:

> ... even if limited only to the cost of replacement power, indicate that the cost to utilities and to their ratepayers will be in the range of tens of millions of dollars a month for each plant ...

Breaking the Vicious Circle

U.S. Supreme Court Justice Stephen Breyer addressed this situation in a book entitled, *Breaking the Vicious Circle: Toward Effective Risk Regulation.* This slim volume contains the Oliver Wendell Holmes lectures Judge Breyer delivered at Harvard in 1992. He describes how interaction among the regulatory agency, the public, and the legislative body involved in regulating a particular hazard creates a vicious circle in which regulatory requirements often proliferate out of control, powered by fear fanned by the regulatory process itself. The problem is exacerbated as each step in the process generally uses the most conservative assumptions. He quotes a study by the U.S. Office of Management and Budget (OMB):

> "Suppose that there are ten independent steps in a risk assessment and prudence dictates assumptions that in each instance result in risk estimates two times the expected value. Such a process would yield a summary risk estimate that is more than 1,000 times higher than the most likely risk estimate. Because there are usually many more than ten steps, and many of them will incorporate conservative biases that exceed [that] order of magnitude, risk estimates based on such practices will often exceed the most likely value by a factor of one million or more.

Having noted, "We regulate only some, not all, of the risk that fills the world," Breyer concludes that excessive concern with any one source of risk necessarily deprives other problems of the attention and resources they deserve.

The Voice of the People

A good political advocacy group will try to convince public officials that it is a grassroots group, that it speaks for "the public," or at least for a good fraction of it. Anti-nuclear advocates have done a good job of this. One result is that the federal government generally dealt with the nuclear industry as an adversary, not only with regard to regulating its safety but also in connection with sales of nuclear plants abroad. American companies like Westinghouse, GE, Bechtel and the like were counting on significant income from building nuclear power plants over seas.

Other nations that competed with us for those sales often had their governments and their financial institutions working hand-in-glove to present attractive financing arrangements we could not meet. This was particularly difficult during the Carter presidency when U.S. companies had to pay 20 percent or more for money due to inflation, and Japanese and European firms were able to offer financing arrangements that sounded like a 1990s car dealer promotion. For an item as expensive as an entire nuclear power plant complex, financing can be a critical factor.

Indeed, most of the cost for a new nuclear power plant generally went to lawyers and bankers, not for materials and technical labor.

So we must ask: What is the public's attitude toward nuclear power? Here again, the facts are different than you might conclude from the media. Granted, many people react negatively to questions like, "Would you want a nuclear plant in your neighborhood." Most people wouldn't want *any* massive industrial facility in their neighborhood. But three separate 1993 polls asked the general public, opinion leaders, and congressional staffers if they thought nuclear energy should play an important role in our country. Sixty-nine percent of the public, 72 percent of the opinion leaders, and 83 percent of congressional staffers thought it should. In contrast, less that half of these nuclear advocates believed that the majority of the public felt as they did.

A more recent study (March 2001), by Bisconti Research with Bruskin Research, showed that this favorable attitude of the public continues to increase. The percentage who agree, "We should definitely build more nuclear energy plants in the future," went from 42 percent in October 1999 to 51 percent in January 2001 to 66 percent in March 2001. On the March survey, 70 percent said nuclear energy should play an "important role" in America's future, and 87 percent agreed that existing plants should have their operating licenses extended.

The media keep referring to this public support of nuclear power as "surprising." I'm always amused by articles that tell us how surprising it is that most Americans do or believe certain things—attend church, favor nuclear power, or vote for people the pollsters had written off. What such statements really tell us is that many reporters and commentators who claim to know what's going on are, in fact, *surprised* by what's going on. We should remember that as we read their solemn pronouncements.

In my youth, there was a Chicago radio program that featured a couple of comedians called Stoopnagle and Budd who had an "expert" they liked to interview. They assured us that this expert "knows more about people, as such, than anyone else in the world, with the possible exception of people themselves." As proud members of this exceptional group—people—we should be prepared to deal with pundits and experts as we deal with everyone else: neither obsequiously nor disdainfully, but with the simple respect due a fellow human who has certain knowledge and experience we don't have, but who probably has no more general wisdom than the rest of us.

There is no doubt that potential builders of nuclear power plants found the atmosphere of government supervision and public participation so burdensome that they would do almost anything else before deciding to build another nuclear plant. I remember a simple example that illustrates the problem: a hearing called to consider a minor design change in an existing plant. As usual, there were about a hundred people in the room: officials and technical personnel from the operating utility company; engineers and scientists from the company supplying the reactor plant; representatives from the architect-engineering firm that built the plant and designed the steam system; engineers and lawyers from the NRC; research personnel from the Department of Energy; many, many consultants working for each of these organizations; engineers and lawyers from various antinuclear groups looking for targets of opportunity; and miscellaneous neighbors, reporters, and unidentified members of the public who might well have been just looking for an air-conditioned place where they could sit down for a while.

The reactor engineer presenting the technical story in this case found that the NRC engineer he had been working with had been deflected off to another problem. The presenter looked around the room, saw another NRC engineer in the crowd, and started addressing the story to him. But soon that engineer was also called out of the meeting, and the presenter continued his presentation, looking about vaguely for someone else to whom he could address his remarks. People were continually leaving the room and others were entering. Numerous informal sub-committees were holding discussions under their breath in various corners of the room.

Suddenly the presenter stopped. "Is anyone here from the NRC?" he asked in a loud voice. The other conversations died down. Several people looked as if they were about to volunteer, but he added, "I mean actual, full-time employees of the NRC. I want to be sure that someone who is going to make the decision on this, or at least recommend a course of action—I want to know where that guy is." There was an embarrassed silence as we all looked around the room and looked back at our shoelaces. "Dammit!" he exploded. "We're talking to ourselves! I'm not saying another word until we get somebody from the NRC back in here to listen." And he sat down. Eventually they found one NRC employee who was willing to listen.

This was a frustrating and hugely expensive process, as the review of each plant took years and tied up scores of high-priced people debating the same questions plant after plant. Having lived through this, I've never been able to take seriously the claim that this regulatory process does not add significantly to the cost of nuclear power. The process is run largely by lawyers, as a legal ritual, despite the fundamentally technical nature of the questions at issue. The pity is that many persons still believe that the nuclear industry is allowed to proceed without adequate public scrutiny. One can disagree with some of the decisions made during this process, but there is really no basis for arguing that the public's need for information and for presenting complaints and suggestions about the nuclear power program is somehow being short-changed.

The Public vs. the Expert

The question of the Public vs. the Expert is a difficult one. Some people view all experts as biased and corruptible and give their advice little credence. Others cower before their expertise, convinced that experts have knowledge and wisdom unattainable by mere mortals. Either of these positions leads to poor decision-making. In the nuclear field, we seem to have found a way to combine the worst features of both these contradictory positions. We are asked to ignore government or corporate "establishment experts," since they are presumably defending a political position and, therefore, cannot be trusted. But we are asked to accept without question the "outside experts," who are presumed to rise from the grassroots to express without bias the views of The People. We don't decide other technical issues this way—the prime interest rate, air traffic control procedures and the like. Yet in nuclear power, we require all comments and suggestions—even those clearly technically ignorant—to be treated with ponderous deliberation.

The cost of these extensive hearings cannot be dismissed by arguing that fat cat corporations can afford it. In the final analysis, corporations don't pay taxes; they merely collect taxes from their customers. We consumers ultimately pay the bill, both for the utility and its helpers and for the salaries of the regulators as well. But there is no mechanism in the present system for taking cost into account. The NRC and all the other decisive players in this game are required by the rules to look only for ways to make a plant safer and safer. The argument that the cost of a change might far outweigh its value is inadmissible evidence, like a confession obtained through torture. And additional "safety features" added to a plant do not necessarily make it safer.

But whatever we might do to clarify and speed up these proceedings, there will always be a number of gray areas where there is not an unequivocal answer. Many people tend to think of science as dealing with absolutes: two and two always make four. Not five, nor even 4.1. Just four, and there can be no arguments about this. But even in the technical domain, there is a difference between facts, opinions and judgments

Facts

A woman once told me that she feared nuclear power, because every power plant was a potential Hiroshima bomb. I told her that was a physical impossibility, that there was no way an American power plant could be detonated like an atomic bomb—the laws of physics would not permit it. She just smiled sweetly and said, "Well, that's *your* opinion." I could not persuade her that there are some matters on which we don't get to hold varying opinions. This is an important point on which any rational debate must rest. But there are people who insist the earth is flat and that the space flights were faked. And there are people who even question the very idea of facts. Nevertheless, in trying to understand any controversial situation, it is still worth trying to distinguish facts from opinions and judgments.

Opinions

Even with the facts agreed upon, there will still be plenty of room left for a good argument. A proposed new power plant—or a new shopping mall, for that matter—will have certain characteristics that are factual. It will employ so many people, it will occupy so much space at a specified location, and so on. But then the opinions come in. Will the people going in and out of the facility create significant traffic congestion? Will the facility overload existing water, sewage, school, or transportation systems? Such questions require us to use all available facts and then make some guesses—that is, form some opinions—as to how these will all work out in practice.

Judgments

The final answer, how you feel about a proposed project, involves more than fact and opinion. You have to feed in your personal values. In your opinion, the traffic congestion caused by the facility might be less severe than others are suggesting. But if you are so opposed to even the slightest congestion, then you might oppose the facility. Or you might feel that the need for jobs in your area is so acute that the employment opportunities offered by the facility will outweigh the other considerations and you would favor it.

If the analysis that you and your neighbors have done is rational and thorough, there should be agreement on the *facts,* some differences of *opinion* as to how these facts will affect the area, and perhaps great differences of *judgment* as to whether one should buck it or back it. In an enlightened situation, further discussion and efforts to bring others around should be a pleasurable intellectual challenge leading to increased mutual respect, rather than the bitter and frustrating wrangles that lead people to the brink of hate and make them despair for the future of the race.

In a rational discussion of nuclear power, it is important to know the reasons your adversaries disagree with you. They may find unconvincing the experimental basis for concluding that low-level radiation is harmless, in which case you might introduce the epidemiological data. But if their opposition is really based on a concern that nuclear power necessarily involves big corporations and big government, then there is no point quibbling over the radiation data. To win debate points, the opponents throw in scare talk about radiation, and the nuclear advocates accuse their opponents of being against progress and against business. As a result, it is not generally recognized that there is little disagreement about the facts between nuclear power advocates and those opponents who have actually studied the technical data.

"Whose Plant Is This?"

The Shippingport plant was the product of many separate organizations, each of which devoted a large number of their best people to it. Rickover's group had overall line responsibility in the AEC, working through its Pittsburgh Area Office, which was under the AEC's Chicago Operations Office. Westinghouse developed the nuclear reactor plant and much of the associated technology. The Duquesne Light Company furnished the site for the entire project and built, operated and maintained a new 100,000-kilowatt turbine-generator plant at no cost to the government, and furnished $5 million worth of services and materials toward construction of the nuclear portion of the plant. Duquesne also agreed at the outset to purchase $11 million worth of steam from the government's nuclear plant over the first five years, to run the turbine and generate the electricity.

Stone & Webster Engineering Corporation was selected from nearly 100 architect-engineer firms to assist Westinghouse in the nuclear plant design for cost plus a one-dollar fee. Burns and Roe, Inc. was selected by Duquesne to build the turbine-generator plant, and Westinghouse selected the Dravo Corporation to install the components and facilities of the nuclear plant, again on a cost-plus-a-dollar basis. Other major companies involved include Combustion Engineering, which built the reactor pressure vessel, and Foster-Wheeler and Babcock & Wilcox, each of which built one of the two steam-generator heat-exchangers. Pittsburgh-Des Moines Steel Company built and erected the mammoth steel containers that house the nuclear plant.

Although there were many fierce debates among these various organizations during the course of construction, the spirit of this pioneering project was so upbeat that each of the organizations felt that, fundamentally, the plant was "really" theirs. On September 6, 1954, President Eisenhower waved a neutron-emitting wand before a radiation detector and remotely started an unmanned bulldozer at the Shippingport site, officially initiating construction. Actual work on the site began the following May. Two and a half years later, on December 23, 1957, Shippingport achieved full power. Electricity was now flowing into Duquesne's commercial grid, supplying the homes, farms and factories of the entire Pittsburgh area. The military atom had finally gone fully civilian.

Starting MPR

By 1964, I had worked for Rickover for fifteen years, the last ten as his Technical Director. Nuclear submarines were being built at several shipyards so fast that we even had two launchings on the same day. Submarines with intercontinental missiles were standing guard under the Arctic icecap, and the mighty two-reactor *Triton* had gone around the world submerged. Nuclear-powered surface ships of three different sizes were in operation, and the first three were preparing for a round-the-world showing of the flag in the manner of Teddy Roosevelt's Great White Fleet.

Under President Eisenhower's Atoms for Peace program, we had built the world's first commercial atomic power plant, and commercial plants based on that technology were springing up like mushrooms after a rainstorm. And I was beginning to get restless.

As it turned out, so were my colleagues Bob Panoff and Harry Mandil. Over a lunchtime snack, we discovered that we were each beginning to wonder what we could accomplish if we didn't have the powerful shadow of Rickover behind us. Could we, on our own, run an engineering company in the commercial world under the same principles of professionalism and technical excellence that Rickover had created and sustained in his program?

Figure 8.5 Seaborg congratulates founders of MPR Associates, Inc. (*DOE*)

We finally decided we had reached the stage of life—our early forties—that we had to find out. We told Rickover our feelings and, after a decent interval of several months to accomplish a proper transition, we left to form an engineering firm we called MPR Associates, Inc. We found that we could run such an operation and there was a real market for engineering services of the highest quality. And, although it was no easier doing this at MPR than it was under Rickover, we found there was a great deal of satisfaction at all levels of the company in achieving this on our own, and in being part of such an organization.

9. Setting Up Procedures to Evaluate Public Safety—from Scratch

As the date of the initial start-up of the *Nautilus* submarine power plant began to emerge from the mists of the future, we spent more and more time thinking about how to get the required safety review and clearance. The radiation levels resulting from daily operation of the plant concerned only the crew. The public was not involved, and outside experts felt we were being remarkably conservative. But for the submarine and subsequent ships to be practical, they would have to be able to go into and out of busy harbors, and there was no precedent anywhere in the world for operating a power reactor in a populous environment. In fact, there were no power reactors anywhere, and even the machinery for judging their safety had not been set up.

One day, Rickover called me in and asked rhetorically, "How about the people of Groton and New London [Connecticut, where the *Nautilus* was being built]? How are they going to react when they suddenly find they have an operating nuclear power plant in their front yard?" Of course, I didn't know any better than he did. "Maybe they'll feel we should have talked with them more ..." he went on, "asked them if they have any questions or concerns. I don't think we've done everything we could."

"Get together with Carl Shugg [manager of the Electric Boat Company]," Rickover continued. "Help him prepare some talks for Rotary Clubs and so forth, tell people what we've done about safety, radiation, and other stuff they might worry about. Get them to ask questions. We've got to get any fears or worries out on the table."

The talks were duly prepared and given. Shugg was an intelligent, open and articulate person, respected in the community. He knew what he was talking about, and he did a good job. The consequences were surprising. “People were bored,” reported Shugg. “They have seen ships go from wood to steel, from rivets to welding, from sail to coal to oil. Their livelihood depends on shipbuilding. They’re used to adapting to new technologies as they come along. For them, nuclear is just one more. They didn’t really have any questions and weren’t much interested in my answers. I don’t think we have a problem.”

“Don’t be too sure,” muttered Rickover. “Things change.”

Half a century later, I often recall this and other such conversations when I am confronted by people asking why we keep trying to cover up the safety problems posed by nuclear power.

The Safeguards Committee

In the late 1940s, when the nuclear energy community first started to face up to the safety questions inherent in nuclear reactors, they looked around for a standard or a model to go by. But there was none. They researched the history of boilermakers, chemical plants, smelters, biological research facilities—every potential public hazard they could think of. The results were sobering. In nineteenth century America, a highway or railroad bridge was failing every other week. From 1825 to 1830—five years—there were 42 recorded boiler explosions aboard American steamboats, each one killing an average of 6 to 7 people. Then in 1830, the boiler on the *Helen McGregor* blew up, killing 50 passengers in one stroke. By the end of the century, boilers of all kinds were exploding in America at the rate of more than one a day. So the Hartford Steam Boiler Inspection and Insurance Company, working with the American Society of Mechanical Engineers, helped bring about legislation and changes in practice that brought the problem under control. There were, however, other problems that had not yet caused public disasters and seemed to be accidents waiting to happen.

D. Allan Bromley’s book, *The President’s Scientists*, quotes a particularly dramatic one:

> On a bright spring morning in 1845, the *Princeton* steamed out into Chesapeake Bay for the initial tests of its 12-inch guns....When [the first gun] exploded, as it did, it killed the Secretary of State, the Secretary of the Navy, a naval captain, a Congressman from Maryland, and ... sundry other dignitaries. Had it not been for the fact that President Tyler had been detained briefly to finish a military ballad below deck, he, too, would surely have been killed. This is the sort of thing that gives technology a bad name!

(Actually, the incident occurred February 28 or 29, 1844, and the big cannon had been previously tested several times. But the relevance of the story here remains unchanged.)

"My God!" said Miles Leverett, Technical Division Head at the Oak Ridge National Laboratory. "Do you realize several good-sized American cities have tons of chlorine gas stored in rusty tanks nearby and upwind? Chlorine! That's what was used as poison gas in World War I. And some of these biological warfare facilities! And the way some industrial wastes are handled—heavy metals organics! And municipal sewage treatment plants—they're run very casually and they have the potential to spread plague like in the Middle Ages. These are no standards for us to go by. We should certainly do a lot better than that with nuclear reactors!"

A simple calculation, with unrealistically pessimistic assumptions for the nuclear case, showed that failure of a single railroad tank car of chlorine near a populated area would produce about the same public hazard as meltdown of the Materials Testing Reactor (MTR) that was being designed for Oak Ridge. In each case, quick temporary evacuation of the area would be required to prevent injury and death to many persons in the vicinity. This became a sort of rough standard of hazard evaluation for nuclear reactors, and the MTR was built at the isolated National Reactor Testing Station in the Idaho desert rather than near Oak Ridge. (Ironically, in 1979, the same year as the accident at Three Mile Island, a wrecked train-load of chlorine in Mississaugua, Ontario, Canada required the evacuation of 200,000 people The incident was all but ignored by the media, and shipments of chlorine to and through populated areas continue.)

So the Atomic Energy Commission set up an Advisory Committee on Reactor Safeguards composed of prominent people not beholden to a nuclear employer. Because of the stature and integrity of the members and their complete independence, this group achieved more and more credibility and respect in the field, and in 1957, Congress passed a bill making the Committee a statutory organization.

In the early 1950s, the Chairman of the Committee was C. Rogers McCullough, whom I knew from our days at Oak Ridge. At Rickover's request, I went to meet with him alone in his office in Saint Louis. In our first conversation, I was not seeking any answers, just posing some questions and asking if he felt that further informal discussions might be useful. He agreed they would, and the next time I brought with me Sidney Krasick, the soft-spoken, highly respected head of physics at the Westinghouse Bettis Laboratory. Krasick was the person primarily responsible at that laboratory for the nuclear design of the reactor and its dynamic and safety characteristics. Both he and McCullough were unusually mature and thoughtful scientists, and I knew such a meeting had the maximum chance of leading to a wise and practical course of action.

Rickover had said, "You can't just go in and ask someone to approve full-power operation of this reactor in a busy port, just like that. We'll take it one small step at a time, and ask approval of just that step. First, we'll ask for approval to run the initial physics tests. The reactor will be critical, but we'll be at essentially zero power. It will be just like running one of those little research reactors that run at only a few watts of power. You don't build up any significant amount of fission products that way. They've already approved several of those research reactors for operation at universities. That's all we'll ask for at first. Then we'll come back with the data from that operation and procedures for the next step, maybe low-power tests."

When we put this plan to McCullough, he listened thoughtfully and responded: "I agree that getting approval for the zero-power tests should not be difficult. But isn't that just putting off the really difficult questions? What do we gain by that?"

Krasick and I explained that this procedure would allow us to approach each next difficult stage with as much information as possible. We would have the technical data gathered in the previous phase, and we would also be getting to know each other. We would be learning how the Committee members think and what they need to satisfy themselves on safety questions. The Committee in turn would be learning firsthand just how intelligent, well-trained and well-disciplined Rickover's sailors and officers actually were in the environment of a nuclear power plant.

McCullough said the idea had appeal, but he wanted to think further on it. He also wanted to talk with other Committee members. We, of course, were in no position to pressure the Committee. No one was. We knew they met in secret and published their findings when they had finally agreed on the wording. None of the members was dependent on nuclear work for income or for reputation. So we thanked McCullough and went home to bide our time.

Rickover had one more idea. He knew that the Committee was worried about how strictly any group of operators would stick to all the various procedures required of them. He knew that some very smart scientists had done some very foolish things in that regard. Rickover also knew that submarine sailors take a very serious view of discipline and the need to follow procedures to the letter, and he wanted the Safeguards Committee to see that for themselves. So he arranged for all thirteen members to ride submarines—diesel-powered, of course; that's all there were. It took three submarines to accommodate all of them, and the crews demonstrated fire drills, collision drills and other emergency procedures. These were ordinary submarine crews, not nuclear trained, and the drills were standard submarine procedures. But they were a marked contrast with how such drills might be carried out by graduate students operating a research reactor. The Committee members were impressed.

In due course, the Committee approved the first step, and at each subsequent meeting, we reviewed the information to date, let the members explore the ship and the prototype plant, and we got approval to take the next careful step. We carried this process through to power operation in port and then refueling. We also had some meetings in which we just brought them up to date on the program and answered questions. In the course of this process we observed another instance of how differently submarine crews and scientists view procedures: The Atomic Energy Commission itself was unwilling to apply Rickover's stern criteria to the Commission's own weapons facilities or its national laboratories. The AEC decided that many of the rules it had set up for others were just too onerous in practice to impose on its own non-naval facilities.

Bringing in the U.S. Public Health Service

Another step that Rickover took early in his program, to assure that no safety measures were overlooked, was to enlist the help of the U.S. Public Health Service. In the late 1950s, some of the states were beginning to consider drawing up legislation of their own to define how radiation and radioactivity should be handled within their borders. Pennsylvania was one of the first of these, since Rickover was building the world's first nuclear power station at Shippingport, near Pittsburgh, to deliver power through the commercial grid of the Duquesne Power and Light Company. The state public health people had sought help from their federal counterparts, and they in turn had come to Rickover and his people, who had most of the relevant experience and information.

I was present when Rickover first met with Dr. James Terrill of the U.S. Public Health Service. Terrill was the kind of person Rickover liked to deal with: he could do something for the program, he was intelligent and agreeable, and he truly believed in the rightness of what Rickover was asking him to do.

"Jim, I really mean it when I say I want our facilities, naval as well as Shippingport, to meet all the same standards that the commercial guys are going to have to meet. I'm not asking for any favors or special treatment. If anybody in my program tells you different, let me know about it, and you won't have to deal with *him* anymore."

"But what do you want from me, Rick?"

"I want you to review and comment on all our procedures and standards that might have public health implications. If you think we have some papers you haven't seen, I'll give them to you. You be the judge of what you need to see. And I want you to assign a career Public Health Officer to each of our major facilities as a full-time position."

The Public Health Service agreed to do this and quickly made available competent career personnel. Later, when reporters asked questions as to whether there might be radioactive contamination when nuclear ships began to operate out of the Mare Island Naval Shipyard near San Francisco, the questions were routinely referred to CHINFO, the Navy's public information arm, which began to prepare a reply.

"You fools!" shouted Rickover when he heard about it. "No words from the Navy will be credible in this sort of situation. The Navy has an axe to grind. Why do you think I put Public Health Officers there? Let me handle this."

He immediately called Terrill and said, "Jim, there are some characters claiming you're not doing your job of protecting the public. Your guy at Mare Island has the details. Get in touch with him and prepare a proper response. Let me know if you need anything from me."

Some people saw the Public Health Service arrangements as the sort of shrewd public relations move for which Rickover was famous. "This is no gimmick," he grumbled. "The guys at CHINFO are chortling as if we had pulled a fast one on the public. The damn fools! They don't understand that the Public Health Service has statutory responsibility for protecting the health of the public; they have the organization, the expertise and the public charter. I'm not playing games here. This is the way it's supposed to be! And I'm not interfering with the public health people doing the job right. In fact, they know I'll be mad if they *don't* do it right. "

This incident sheds a little light on the apparent contradictions in Rickover's thoughts and actions. Sometimes he was seen as the consummate light-footed politician, quick to use any legerdemain or corner-cutting for expedience. Other times he surprised people by being a stickler for the letter of the law. But his thought processes in these situations were usually solid and consistent if not always obvious to outsiders. He was a student of history and of government, and he understood the intent and the history of important legal concepts. For example, when British naval officials requested, ever so politely, that they would like permission to routinely come aboard our nuclear submarines as they entered British ports, "to welcome you ashore and have a spot of tea," our submarine captains could see no harm in such a pleasant social ritual.

"You damn fools!" roared Rickover when he found out. "We fought a war in 1812 over just that issue. The U.S. has never conceded the right of a foreign inspection party to board our military vessels. You clowns have just set American foreign policy back 150 years." Of course our officers protested to Rickover that these were not inspection parties—just friendly social visits. "Yeah," said Rickover. "And what do you think will happen to the first skipper who says he doesn't want one of these friendly social visits? What then, eh? Don't you think they'll press him to find out what he's trying to hide? Damn it! You guys have now established a precedent."

Similarly, when the captain of one of the first visiting submarines was asked if he was bringing in any nuclear weapons, he was relieved to be able to answer unequivocally, "No!" Again, Rickover had to point out that the captain had just created a problem for the first ship that *did* have nuclear weapons aboard. The only proper answer in all cases is, "It is U.S. policy never to confirm or deny the presence of nuclear weapons aboard its warships."

The Nature of Safety

This book is about power plants, not about weapons facilities. I am not knowledgeable about weapons facilities, and they operate behind a screen, which precludes public scrutiny. In America and most other countries, nuclear power plants, unlike weapons facilities, are built under unprecedented public review, and all of the relevant facts are out on the table. We do not have to take any government or corporate official's word for anything. We can judge the facts for ourselves. I join with Richard Rhodes, Pulitzer Prize-winning author of The Making of the Atomic Bomb, *who urged the public to "distinguish the continuing success of civilian nuclear waste management from the appalling mess that half a century of weapons production has left behind." That statement applies to nuclear ordnance, not nuclear propulsion. As noted here, the naval nuclear propulsion program chose to meet all civilian radiation safety standards and requirements.*

One of the fundamental aspects of safety—that is, the safety of a complex system like a power plant—is that it is a *resultant* attribute, not an independent one. Independent attributes are such characteristics as having stainless steel piping, being able to produce a particular amount of power, and having one, two, or four piping loops circulating the cooling water. These are characteristics that you can decide on and then incorporate in the plant. They will affect other characteristics of the plant, but you can generally have them if you want them.

Safety, on the other hand, results from many factors: the plant's physical characteristics, the selection and training of the operators, the quality of the operating procedures, and the quality of the materials and workmanship that went into building, maintaining and operating the plant. You can't say, *Let's put 10 percent more safety into this plant,* any more than you can decide to be 10 percent healthier tomorrow, or 10 percent happier. Safety is really a measure of the health or basic happiness of a plant, and trying to grab onto safety and keep it is like trying to grab and hold onto the bluebird of happiness.

This sounds obvious when you state it so crudely, but it is frequently viewed very differently. In many industries, safety is often treated as if it were an independent characteristic that could be pushed into, or pasted onto, a plant. In these cases, all the focus is on quick amelioration of symptoms with little thought for what effect the fix will have on the overall system in the long term. In the nuclear industry, the regulators do, in fact, put a huge amount of time and effort into trying to ensure that the fix does not create any new safety problems. But treating safety as a factor independent of other desirable characteristics such as reliability, ease of operation, and ease of maintenance, leads to some situations that are bizarre in the extreme.

Singling out safety as a predominant priority, overriding all other considerations, can ultimately weaken everything, including safety itself. It is akin to a nation focusing so strongly on "national security" that it eventually leads to environmental degradation, loss of civil liberties and economic collapse—the very antithesis of national security.

Some Regulatory Nightmares

Let me give some examples of the kind of problem that can arise when narrowly-defined safety is given overriding priority, for instance, protecting against a saboteur. Some safety specialists concentrate on that issue and urge that the plant be designed so that a potential saboteur, who might have eluded the security forces, can have access only to one or two small portions of the plant, the others being in separately walled-off compartments. Then the question arises: Who has keys to each of these compartments? To be effective, the system should ensure that only a few persons have access to the keys. These design requirements can be met, and they have been. Now suppose, early one Sunday morning, there is a water leak, or a valve malfunction, or an instrument acting up, that requires immediate access to a certain compartment. Phoning the supervisor who has the key yields no response, for he is away on a weekend fishing trip. The backup keyholder is home, but her line is busy. She has been trying to reach a doctor for her sick daughter, but the doctor's teenager is on an interminable phone call to his girlfriend. We hope that eventually a keyholder will be found and can be persuaded to come in. But surely overall safety has not been served by this over-reaction to one small part of the safety problem.

In another case, a regulatory engineer has suggested a cumbersome solution to a problem that has been holding up approval of the plant. If the plant designer agrees to go along, it looks as if approval may be expedited. The senior vice-president of the company building the plant hurriedly huddles with his engineers.

"Can you do this? How long would it take? What would it cost?"

The engineers reply with a frown. "Sure, we could add all that junk. It would delay us several weeks and cost several hundred thousand dollars. But we'd still get approval faster than trying to talk this character out of his bright idea."

"Would it work? Would it do any harm?"

"Yeah, it'd probably work all right, and I don't see how it could cause any problems. But it's just a lot of unnecessary garbage."

The VP knows that hundreds of millions of dollars have been borrowed to build the plant, and he is not permitted to get any of it back until the plant is actually selling electricity. (In this respect, nuclear regulations in capitalist America are tougher on utility companies than in Britain, where even the socialized British industries were allowed to charge construction costs as they are incurred.) If this particular plant was undergoing its safety review in the late 1970s, as many were, the borrowed money might be at 20 percent interest or more, and the delay would, therefore, be costing over half a million dollars each day for interest alone. In this situation, a company is under great pressure to go along with any action that might speed up the years of safety review, regardless of its technical merit, so long as it didn't actually make the plant unsafe.

Gray-Zone Decisions

In many other cases, safety questions arise that can be argued either way. For example, does adding additional backup electric power generators increase plant safety? All nuclear plants are required to have one or more extra emergency electric power supplies coming in on alternative lines from the main power grid outside the plant. There must also be an emergency electric generating system on site, powered by a diesel engine. OK so far. Now the question arises: what if the diesel engine won't start in a hurry? (Sometimes diesels are balky.) No problem, we can add another. Most plants now have at least two dedicated generators per reactor, which serve no other purpose. (If they served any other purpose, then the regulators would not count them as emergency systems.) A plant with two reactors would thus have four or more emergency diesel engines. Then someone argues that the electrical switchgear that hooks up the plant to the operating diesel and disconnects it from the non-operating diesel (or diesels) is, in fact, less reliable than the diesels themselves. Operating statistics show that a single diesel may be more reliable than two or three diesels and their necessary electrical switchgear. And as a separate question we might ask: are systems that are never run except in emergencies apt to be more reliable than equipment that is used frequently? In some cases, frequent testing burned out some diesels, making them less reliable when actually needed—a *decrease* in safety.

The point of all this is not to argue for or against any particular safety provision or to argue for or against the safety aspects of nuclear power plants in general. The point is that labeling something a "safety device" doesn't guarantee it will make a plant safer. It may well make the plant *less* safe.

The Reactor Plant Design Process

There are a number of activities that go on in parallel during the design of a reactor plant, and nearly all of these affect the evaluation of safety in one way or another. By 1950, physicists at Oak Ridge and at the Argonne laboratory in Chicago had worked out the basic techniques for designing the physics aspects of a simple nuclear reactor. Mechanical engineers were beginning to work on the mechanical aspects, heat transfer specialists on the thermal characteristics, metallurgists on the material properties, systems designers on the plant parameters, along with electrical, electronic, chemical and all the other myriad skills that go into a plant design. Each of these efforts had both a theoretical and an experimental component. The theoretical physicists kept devising better formulas, working with hand calculations, aided in later years by computer modeling.

Their experimental colleagues kept trying to get better measurements of nuclear characteristics of the fuel and structural materials, and made nuclear mock-ups of the reactor, called "critical experiments." These experiments started out as simple strips of plastic containing varying amounts of uranium (usually as uranium oxide) and strips of zirconium, in open tanks of water. The aim was to determine not only how much uranium we would need to start up, but how much the maximum power peak might be above the average power level, and how effective each control rod would be, and how the buildup of fission products might change things, and dozens of other physics questions.

It's hard to believe how primitive our tools were in those days. The first physics calculations were done on standard office adding machines, those heavy green desktop mechanical gadgets with whirring gears that accountants used for balancing ledgers. Later we got PDP calculators with punched cards, and then IBM-650s and then 704 mainframe computers—the real thing! Then we tried the Philco-2000, then Control Data's 6600 and then their 7600. From the mid-1950s to 1970, our program was a major driving force in the world for large-scale general purpose computer development.

After we had used a given computer for a couple of years, we had to start planning for the next generation of computers, because the capabilities were improving so fast (even as now). But in those days, you had to completely rewrite all of your computer programs with each change because there was absolutely no compatibility from one computer to the next, and the rewriting took nearly two years. Initially we did this on special computer programmers' paper, writing everything in machine language, that is, in long rows of ones and zeros in little printed boxes.

We engineers mostly stayed away from such gadgets, preferring to depend on our slide-rules, or for very precise work we used books of five-place logarithm tables. (Logarithms are exponents, and sliderules had logarithms etched on them. By aligning the markings, you were effectively adding exponents, which was the same as multiplying the base numbers. Readers born since the war will have to get their grandparents to explain log tables and the sliderules derived from them. It was a whole different world.)*

---

*The kids might get it more easily if you first tell them the old story about Noah releasing the animals after the flood and ordering them to go forth and multiply. The snakes had been kept in the soggy bilges for forty days and forty nights and were in pretty sad shape. So Noah went out and cut down some trees and built them a little table that provided a dry platform and caught the sunlight. Noah put the snakes on it to dry out. Soon they felt better and they went out to join the other animals, showing that *with the help of log tables even adders can multiply.*

The experimental world was equally primitive. To study radiochemistry, we bombarded water with atomic particles from Van de Graaf machines, the old "atom smashers" featured lovingly in *Popular Science* and in mad scientist movies of the 1950s. Heat transfer tests were run with electric resistance-rod heaters, the same kind of heaters used in electric stoves at home. To test the dynamics of the power plant control system, Herb Estrada, Nelson Grace and others at Westinghouse made an electric analog of the system.

They worked out mathematical equations that described how the plant would react to various changes in temperature, control rod position, power demand, etc., and then built electrical networks that followed the same equations. Thus, they could scan the electrical network voltage readings that corresponded to temperature, or power, or whatever they had set it up to simulate.

In later years, they had parts of the system modeled with digital devices, which is a very different way of analyzing systems. But they were able to get the best of both worlds with their analog-digital hybrid. Today we would have trouble dealing with such a device, but it was effective in getting the answers we needed, when we had no experience with real operating nuclear power plants.

I used to argue with Rickover about how to proceed conservatively in this situation. "After you guys figure out exactly how much uranium to put in, I want to add a little more," he would say. "You may be exactly right. I know you're all goddam geniuses. And I know uranium is scarce. But humor me. I just want to be safe. I don't want to build this whole plant, pull out all the control rods and find it won't start up, just because you guys were figuring it a little too close."

I would protest, "Captain, I want to be safe, too. Starting up is only half the problem. I want to be sure it will also shut down. We're working between pretty tight limits." Of course, I wasn't the person doing the actual criticality calculations; but Rickover would always argue with anyone who might give him a different input.

Providing Safety in the Next Generation Power Plants

In the lull following the Three Mile Island accident in 1979, a number of novel plant concepts were suggested that claimed to offer improved safety. Regulators and other safety professionals tend to keep focusing on "fixing" symptoms, so design features are explored that offer solutions to ever more esoteric hypothetical safety problems. For example, the entire power plant could be buried underground on the premise that failure of the containment boundary would then present no problem. It is not clear, however, that leakage into the earth is any easier to deal with than leakage into the air, and the ability to inspect and repair the containment boundary is more difficult underground. In addition, corrosion of an underground containment boundary is apt to be more severe. Other, radically different, design concepts call themselves "inherently safe," but they rely on complex, untested hydrodynamic phenomena, rather than using the simpler, more maintainable, more reliable systems on which much experience has been gained, both in nuclear plants and in many other systems.

But the mainstream of the nuclear community emphasized the so-called evolutionary approach, as opposed to the revolutionary or radical concepts. This approach goes back to relating safety to simplicity, reliability and increased safety margin. Starting with current design concepts, we peel off a lot of safety gadgets that have been pasted onto plants over the past forty years, and look at ameliorating the problems that necessitated these gadgets.

First, we downrate the reactor itself; that is, if the plant is designed for 1000 megawatts of output, we put in a reactor capable of generating 1200 megawatts, and never run it above 1000. (This is pure heresy to salesmen and to many engineers, particularly those with no operating experience.) It's amazing how much we gain from this. Many safety shutdown devices and controls can now operate in a less hair-triggered manner, with less sensitive input requirements and more deliberate time response. Some other protective systems are no longer needed at all.

Then we see how many valves, instruments and other devices can be standardized. Previously, each set of valves and devices was separately bid by different purchasing agents. They sent out at least three requests for bids and tended to pick the cheapest one (although they did try to eliminate any that they thought wouldn't really measure up). The result was that each plant ended up with thousands of gadgets, each a little different, each with its own operating, maintenance and test manuals and its own spare parts inventory.

By looking at the plant holistically, instead of as a series of separate parts, we find that many of these devices could be from the same manufacturer, and some could even be the same size, with the same model number, spare parts, and manuals defining procedures for operation, maintenance, test and emergencies. The load on the operators and maintenance people is thereby drastically reduced, and the probability of error consequently reduced. Training, parts inventory and control, and many other aspects of operation are similarly eased.

A class of next-generation, evolutionary reactor plants designated ALWR (Advanced Light Water Reactors) has been developed by each of the reactor makers. A measure of the success in achieving the above objectives is shown by a few statistics (taken from the Westinghouse AP600 plant design, as an example). The following *percent reduction* in number of major components was achieved: Valves, 60%; Pumps, 35%; Length of safety grade pipe, 80%; Number of heat exchangers, 80%; Volume of seismic buildings, 45%; and Length of electrical cable, 70%.

Figures are not available for the reduction in the number of *sizes* and number of *vendors* of components, but these have also been markedly reduced.

Operator Error vs. Design Deficiency

There is considerable evidence that more accidents and incidents have been judged to result from operator error than from design deficiency, but this conclusion raises some questions. To start, note that it is designers and analysts, not operators, who are most often involved in analyzing casualty reports, and designers tend to believe that the operators should have done better. I once saw a report that said that an operator error is one hundred times more likely to occur than an equipment failure, and thus we should add a lot more gadgets. It apparently didn't occur to that writer that one reason there were so many operator errors is that there were so many gadgets to deal with.

There is fundamental merit in the idea that plants can be made safer by making them simpler. Operators were given a lot of blame for the Three Mile Island accident. The operating logs showed that over a hundred alarms and warning lights came on during the first few minutes of the accident—obviously more than anyone can properly deal with, particularly at 4 A.M. (when the crisis began). Yet many designers and analysts reviewing this information concluded that *more* information and warnings should have been available to the operators. I get into this in more detail when we discuss "Learning from Three Mile Island."

The Role of Economics and Politics

The reason downrating and simplification have not been given greater emphasis in various safety programs is primarily a regulatory one. Present U.S. rules require a utility to buy the lowest-price plant or be prepared for a long argument with the rate regulators. This forces plant designers to design the smallest plant, with minimum material and construction costs. If maintenance costs soar because of access difficulties, and operations are complicated by hair-trigger controls, these costs will come out of operating expenses, a different budget at a later time, which is easier to justify to the regulators.

This situation does not lead to the best designs, from either a technical or an economic standpoint. (If it were your own money, you wouldn't do it that way. You don't pick a brain surgeon on the basis of a low bid. And you don't always buy the cheapest car or the cheapest meal in a restaurant.) There is talk of changing this situation, but so far I have not seen much improvement.

The point of all this is that great and real advantages accrue from not looking at safety, or cost, or any other single aspect of a plant as a separate factor to be optimized, but instead from looking at the plant holistically and noting how such factors interrelate. Safety and cost and ease of maintenance may all be improved if we can resist the argument that any of these individual attributes is so important that it must be optimized independently. If you still feel compelled to remark, *Safety is so important; how can anything else be traded for it?* I would reply that we are finding that each of these individual factors improves if the holistic analysis is done right. Any accident or malfunction in a nuclear plant can be very expensive, so the owner can afford to spend considerable money to get a plant that is safe and reliable, year after year.

The Profession of Risk Evaluation

It is not possible to avoid risk altogether. Any source of energy, by its very nature, presents hazards. Coal miners die, and tens of thousands of other people die each year from air polluted by the stack gases of coal-burning power plants. Oil tankers wreak environmental damage and even gas-fired power plants create global warming gases and exploding pipelines. Perhaps the most lethal course of all is not to have adequate supplies of energy—power to provide bright street lighting in high-crime neighborhoods; reliable emergency power for hospitals and elevators; heat and air conditioning in crowded living quarters; power for water treatment, mass transit and sewage processing facilities.

Calculating the risk associated with various activities has become a fairly standardized process, and there is pretty good agreement among people of different persuasions as to how to do it and what the answers are. But the answers are no better than the data you put in. What you do is to take the total number of people assumed killed by the activity in question (for example, use of personal automobiles), and divide it by the number of such events (for example, passenger miles traveled). The number of Americans killed in cars each year is about 40,000, and the number of passenger miles traveled is about four million-million miles (or 4×10^{12}). Dividing 4×10^{4} by 4×10^{12} gives a risk of being killed in an automobile of 10^{-8} per mile, or one death every hundred million miles. If you travel 10,000 miles each year, that comes to 10^{8} divided by 10^{4}, or one chance in 10,000 that you will get killed in an automobile in any given year. If you drive for fifty years, your chance of being killed in an automobile is one in 200. It's not fancy math; it's just simple arithmetic—multiplication and division.

Truth vs. Reality

The arithmetic is valid, but the question is: How real are the answers compared with other risks we face? When we talk about deaths and injuries from automobile accidents, we know we are dealing with something real. We all have close friends and relatives who embody the statistics. This is also true for cancer. But when we talk about deaths from power plants, the problem is more ephemeral. Do you know anyone killed by a power plant? Yet the U.S. Department of Energy's former Office of Health and Environmental Research estimates that air pollution causes 100,000 deaths per year in the U.S. from respiratory illness. This means that one out of every thirty Americans' deaths is the result of air pollution, even though none of these individuals gets labeled in the morgue CAUSE OF DEATH: AIR POLLUTION. Based on EPA estimates, it appears that nearly a third of these air pollution deaths are from pollution caused by fossil-fueled power plants.

Some of these numbers are relatively noncontroversial. For example, death rates for construction workers handling steel, pouring concrete, installing roofing, etc. are known for the entire construction industry. It is usually assumed that people building nuclear power plants, where no special hazards exist prior to operation, will approximately match the national averages for other construction workers. Other numbers, such as the number of people killed from a nuclear accident and the probability of a nuclear accident, are more open to conjecture. The only nuclear power accident we've had in America killed no one and injured no one, and this is after we've run several thousand reactor-years of operation (a *reactor-year* being the operation of one reactor for one year, or two reactors for a half-year, etc.).

Risks from Nuclear Power

Professor Bernard Cohen, in his book *The Nuclear Energy Option* (Plenum Press, 1990), goes a step further in trying to make these calculated risks real and understandable. Since we all die eventually, he expresses risks in terms of *Loss of Life Expectancy* (LLE), the average amount by which one's life is shortened by the risk under consideration.

This is a more complicated calculation, but it makes the risk easier to visualize and to compare with other risks. He lists some typical risks and the number of days a person's life might be shortened through the risk:

Cigarettes (1 pack a day)	2,300 days
Heart disease*	2,100
Working as a coal miner	1,100
Cancer*	980
Being 30 pounds overweight	900
Stroke*	520
Vietnam army service	400
Alcohol*	230
Motor vehicle accidents	180
Air pollution*	80
Occupational accidents	74
Natural radon in homes*	35
Coffee, 2 ½ cups a day	26
Living near nuclear power plant	0.4
All electricity nuclear*	0.04

Figure 9.1 Loss of Life Expectancy

The *asterisk* indicates an average risk for the whole U.S. population; non-marked risks apply only to those engaging in the listed activity. The last line is the risk to the whole population if all U.S. electricity were generated by nuclear power, as calculated by the Nuclear Regulatory Commission. This last number is not based on the actual health effects of the Three Mile Island accident (which were essentially zero), but upon the pessimistic assumptions that even a single gamma ray poses a cancer risk. If instead we use the figures published by a leading anti-nuclear group, the Union of Concerned Scientists, the number, based on even more pessimistic assumptions, would be 1.5 days, still quite a small risk.

Cohen sums it up as follows in *The Nuclear Energy Option*, pp. 130-131):

> Having a large nuclear power program in the U.S. would give the same risk to the average American as a regular smoker indulging in one extra cigarette every 15 years, as an overweight person increasing his or her weight by 0.012 ounces, or as raising the U.S. highway speed limit from 55 to 55.006 miles per hour, and it is 2,000 times less risky than switching from midsize to small cars. If you do not trust establishment scientists and prefer to accept the estimates of the [anti-nuclear] Union of Concerned Scientists, [then these risks of nuclear power become equivalent to] a regular smoker smoking one extra cigarette every 3 months, or of an overweight person increasing his weight by 0.8 ounce, or of raising the speed limit from 55 to 55.4 miles per hour, and it would still be 30 times less risky than switching from midsize to small cars.

If you are thinking of switching to a small car in order to conserve fossil fuels and reduce air pollution, you can accomplish the same purpose with considerably less risk by driving a large electric car whose battery is charged with electricity from a nuclear power plant.

The Rasmussen Report

Several years before the accident at Three Mile Island, the Atomic Energy Commission issued a safety study that was widely discussed in public debate. This report, known as "The Rasmussen Report" after its main author, estimated that some sort of fuel-melting accident might occur to a reactor once every 20,000 years. With 500 reactors operating worldwide, we should expect such an accident every 40 years (that is, 20,000 divided by 500). And in the last 40-plus years we've had one. That's a far cry from the critics' claim that TMI was "the accident they told us could never happen."

No responsible person ever claimed that nuclear power (or any other system for generating power) was without risk. Moreover, we've learned a lot since TMI, and I think it's reasonable to conclude that the chances of another major nuclear power accident, and the consequences of such an accident, are now even less.

Handling Radioactive Wastes

There are three aspects to nuclear power plant safety. We have discussed shielding against the direct radiation from the operating plant, and the possibility of a fuel-melting accident. The third factor is handling the radioactive wastes resulting from normal operations and maintenance.

We in the nuclear power community have always considered that one of the major advantages of nuclear power is that its waste-handling problem is trivial compared to other ways of producing power. Today, many people consider such a statement preposterous. But the fact is that the *quantity* of wastes produced by nuclear power is millions of times smaller than other processes we are used to dealing with. And safely handling small quantities of material, no matter how toxic, is just not a difficult job, compared, for example, to dealing with the millions-of-times-greater quantities of toxic waste produced by a coal plant of the same capacity.

Radioactivity

The *curie* defines the amount of radiation that will stream out of a radioactive source. The radioactivity might be associated with a tiny piece of highly radioactive material or with tons of slightly radioactive material. Your radiation *dose* depends on how close you stand to it and how long you stay there.

The metric equivalent of the curie is called the *Becquerel* (Bq), after Henri Becquerel, the French physicist who discovered radioactivity. One becquerel is an amount of radioactive material that gives off only *one* ray or particle per second. It is an extremely small unit. To give you an idea how small, let us consider salad oil, whose natural radioactivity is about 5 millionths of a curie per quart. (This is about 20 times more radioactive than typical drinking water.) Five millionths of a curie sounds small, and it is. But it is nearly 200,000 Bq. That is, a one-quart jar of salad oil gives off 200,000 particles or rays of radiation every second. That sounds scary, but it is still only pure, unadulterated salad oil. Many of the spills of slightly radioactive water that are headlined by the media as major environmental disasters result from water less radioactive than salad oil. But they sure involve a lot of becquerels.

The radioactivity of a substance is only part of the answer to how hazardous it is. The other part is the pathway—how this material could be inhaled or eaten to irradiate someone. If there is no feasible pathway, there is no hazard.

Lord Marshall, former head of the British Atomic Energy Authority, gives a provocative analogy: Any healthy young man produces enough sperm to impregnate every woman of childbearing age in the Western Hemisphere. On the basis of that figure alone, any one young man is a real danger to the world. But when you take into account what has to happen to bring this danger about, we conclude that the probability of one man's fathering more than one hundred million children can be safely ignored. Similarly, plutonium is a hazard only if it is dispersed in very fine particles and inhaled. It can be safely handled with only gloves for shielding, and even if swallowed is only mildly toxic (comparable, spoonful for spoonful, to caffeine). Sitting on a table or stored in a warehouse, plutonium's latent toxicity is as harmless as the young man's unleashed sperm.

Marshall notes that each person's share of nuclear waste, assuming all his or her electricity is produced by nuclear power for a lifetime, would add up to five gallons of low-level waste, a little more than a gallon of intermediate waste, and about half a cup of high-level waste. Adding an allowance for packaging and shielding, you could still literally store it in a corner of your basement. He then asks: if you would prefer something else, you can get your electricity from coal, noting that oil and gas will soon run out. Your share of the waste from the coal plant would be about 700 cubic feet of coal ash, whose radioactivity alone is significant.

In addition, each coal-fired power plant produces 15 tons per *minute* of the global warming gas carbon dioxide, as much nitrogen oxides as 200,000 automobiles, several tons of particulates each *hour*, and various organic carcinogens and radon gas, all of which are dispersed into the air with the stack gases from other coal-burning plants to cause 30,000 American deaths per year. Since the wastes from a coal-fired plant are five million times greater in weight and billions of times greater in volume than the wastes from a nuclear plant of equal power output, there is no way such huge quantities can be packaged or otherwise kept from harming people and the environment.

Lord Marshall was once called to task by the media for dumping 400 kilograms of uranium into the Irish Sea during the previous year. Although this was done in conformity with regulations, the press claimed such a hazardous practice should be preceded by a public hearing. Marshall confessed before Parliament that he had done as charged. But he went on to note that the coal-burning power stations under his authority dumped *2,000 kilograms of uranium* into the same sea *every week,* but "we do not call it radioactive waste, we call it coal ash."

The French, German, British, Japanese and other nations dependent on nuclear power have demonstrated that the wastes can be fused into a glass or other insoluble form and then put into drums and stored above or below ground. It is virtually impossible for any significant quantity of radioactivity to escape from such a material, as shown by studies of Egyptian and other ancient glasses. By contrast other wastes such as sewage sludge and coal burning by-products are produced in such quantities that they cannot be effectively controlled, and they do in fact repeatedly cause serious public health problems.

Fuel Elements and Fission Product Waste

Most of the so-called nuclear waste from commercial power plants is actually spent fuel elements in which only three percent of the fuel has been used up. It is a valuable energy resource. Most of us in the nuclear business have always assumed that these spent fuel elements would be recycled to recover the unused fuel, and the leftover fission products would be processed to recover other valuable elements. The rest of the material would be disposed of as waste.

This system of recycle and recovery would ensure that our nuclear fuel resources would last as far as we can see into the future, otherwise, they might run out during the next century. We actually had three fuel reprocessing plants operating in the U.S. starting in the late 1960s. But when Jimmy Carter became President, he made a move that complicated the waste-handling situation. He realized that operating the reprocessing plants would result in the production of nuclear fuel, and decided that if this material got into the wrong hands, we might end up with terrorists with atomic bombs. This is what they called the "proliferation problem," the possibility of nuclear materials proliferating beyond the governments now controlling them.

The Proliferation Problem

John McPhee, a careful and non-sensationalist author, wrote a book about the danger of terrorists making atomic bombs from material stolen from private industry. "What will happen," he asked, "when the explosions come—when a part of New York or Cairo or Adelaide has been hollowed out by a device in the kiloton range?" He quoted atomic bomb experts and think tank analysts saying things such as, "I think we have to live with the expectation that once every four or five years a nuclear explosion will take place and kill a lot of people," and "I can imagine a rash of these things happening. I can imagine—in the worst situation—hundreds of explosions a year." That was in 1973, an entire human generation ago, and this has not happened even once, anywhere in the real world. Of course, no one can say that it can *never* happen, but we should put the hazard in proper perspective.

Nuclear bomb experts agree that if terrorists were somehow able to get their hands on enough material to make a nuclear explosive, creating the device itself without complex and highly sophisticated weapons fabrication facilities would be very difficult. It is highly likely that a premature explosion would kill them in the attempt. We learned in the 1990s from the terrorist bombings of the World Trade Center and the federal office building in Oklahoma City that terrorists don't need fissionable material. It is all too easy to get a devastating explosion by mixing a ton or two of ammonium nitrate fertilizer with some fuel oil, putting it in drums in the back of a van and detonating it with a quarter-stick of dynamite. These materials are readily available to anyone. And then we learned that our own airliners can be used for this purpose.

It is important to note that *not one* of the ten or more nations with atomic weapons capability (U.S., Russia, U.K., France, China, India, South Africa, Israel, Pakistan, Sweden, and perhaps North Korea, Iraq, Brazil, Argentina, and others) achieved weapons capability through atomic power plants. That's just not the easy way to do it. This is an important fact. If nuclear power had been outlawed from the start, this would not have deterred any of these nations from making atomic bombs. They all did it the way we did: they built their own uranium-enrichment or plutonium production facilities for that purpose.

Despite this certainty, Carter believed it would be prudent to shut down our nuclear fuel reprocessing facilities and renege on our contractual commitment to process fuel for other countries. Almost overnight, America went from having a near monopoly as supplier of enriched uranium, to being just one more salesman in a competitive international market as other nations set up their own reprocessing capabilities.

Great Britain, France, Germany, Japan, India, and China all developed operating reprocessing facilities and may have even larger facilities in the near future. Belgium, Italy, Switzerland and the Netherlands all count on use of the British or French facilities. Clearly, nuclear proliferation has changed from a national issue to an international one.

As a result of the end of the cold war, we are now faced with the sudden availability of tons of plutonium and enriched uranium from dismantled weapons—a priceless energy resource, equivalent to suddenly discovering another mid-east oil pool within our own borders, already pumped and refined.

Critics who once derided nuclear power because they believed that our uranium resources would not last are now arguing that we should destroy this plutonium because we have more nuclear fuel than we need. For half a century we have cowered before the thought of terrorists stealing plutonium—which has never happened—while gas pipelines and storage facilities explode, oil tankers pollute our shores, and coal destroys our lungs and our landscape. There are no risk-free options.

The Effects of Not Recycling Fuel

The effect on waste disposal of not recycling fuel is this: When used nuclear fuel is reprocessed, the long-lived chemical elements known as *actinides*, which include uranium and plutonium, are removed. The fission products left behind as waste are comparatively short-lived. After about 300 to 600 years, its radioactivity is comparable to the natural uranium ore from whence it came. At that time, you could *eat a pound of the stuff* and it would only double your chance of getting cancer during the following thirty years. But without reprocessing, the actinides remain in place, giving the "waste" a half-life of tens of thousands of years.

Not only do the spent fuel elements contain valuable nuclear fuel, but the fission products left behind after processing contain a number of other rare and valuable elements produced by the fission process: refractory metals such as tantalum and tungsten; noble metals such as platinum, palladium and rhodium; and a variety of rare earth metals and other elements whose use has scarcely been explored because they were previously thought to be in short supply. It has been estimated that the value of such materials might more than compensate for the cost of recovering them.

In addition, the use of radioactive isotopes in medicine and industry has become a $250 billion a year business, accounting for four percent of the U.S. Gross Domestic Product, 3.7 million jobs, $11 billion in corporate profit, and $45 billion in tax revenues. When we are being urged to "recycle, don't waste" in every other sphere of our lives, it is strange to see people urging us to waste and not recycle this valuable material. Ironically, as our nuclear regulatory requirements have grown more onerous, we now have to import most of the radioisotopes used in medical and industrial processes: diagnostic procedures involving trace quantities of particular radioelements, therapies utilizing radioactive materials that seek out particular organs, such as radioiodine for thyroid problems and heavy metals for bone cancer. We also import radiocobalt for industrial radiography of large-scale welds. America, which developed these procedures and pioneered in their use, now buys its radioisotopes from Canada, England, France, Russia and even Argentina.

In any event, the problem of putting a few tons of radioactive material safely into storage is technically trivial. The real "nuclear waste," as Professor Cohen points out, is the money we spend on the commitment to bury this material in some undefined geological formation that can be guaranteed to remain intact for millions of years.

There are many other toxic materials with infinite half-lives—that is, they are not radioactive and their toxicity never diminishes. These include elements such as lead, chromium, mercury, arsenic, antimony, cadmium, and selenium, and numerous commercial compounds, some of which are more toxic, ounce for ounce, than plutonium. Many of these are handled with casual abandon. We can and should handle nuclear waste carefully, fix it in a solid form so that it will not dissolve in water, and then handle it like any other normal waste. It is not a difficult problem.

The Myth of Nuclear Uniqueness

It is widely believed that nuclear waste poses an unprecedented public hazard. Critics like to quote the figure—originally calculated by nuclear engineers—that if all the electricity produced in the U.S. came from nuclear plants, the radioactivity produced as a by-product could kill ten billion people. You don't have to say that very loud to terrify a lot of people. But is it unprecedented?

Let's look at some other substances we produce without causing much trouble. *Barium* compounds find many uses, as rat poison and in fireworks, and in numerous metallurgical and electronic applications. *Hydrogen cyanide* is used as a fumigant and in the production of plastics, fibers, dyes, and various metallographic processes. *Ammonia* is an important fertilizer and a basis for many chemical products. *Phosgene* is involved in the production of many organic compounds, dyestuffs, and resins. *Chlorine* is widely used for water purification and as a basic raw material for the chemical industry.

Following are a sampling of extreme case mortality rates that could potentially result from the annual U.S. production of these five chemicals:

Chemical	Number of Human Deaths
barium	100 billion people
hydrogen cyanide	6 trillion
ammonia	6 trillion
phosgene	20 trillion
chlorine gas	400 trillion

Figure 9.2 Production of Toxic Materials

These numbers, including the nuclear waste figure, are calculated by dividing the estimated lethal dose into the total quantity of material produced. They are purely hypothetical numbers. A large public swimming pool may contain enough water to drown a million people, but we don't think of it as a potential mass killer.

When Ralph Nader stated in a public debate that a pound of plutonium could kill every person on earth, he was wrong by a factor of one thousand or more. Ignoring that discrepancy, his debating opponent, the nuclear pioneer Ralph Lapp went to the heart of the matter by responding that the same was true of a pound of fresh air. In each case, death would occur only if just the right amount of "poison" was injected into just the right place in the body. A small bubble of air in the bloodstream can kill just as effectively as a lethal injection of plutonium. Of course it would be ludicrous to think of fresh air as "toxic." But Nader's statement about plutonium is just as ludicrous, for exactly the same reason. The fact is although we have handled tons of plutonium in America for nearly half a century, it has never caused a single death. Not one.

Even anti-nuclear activist Sheldon Novick of the Sierra Club, whom I quoted at the beginning of Chapter 7, wrote in his 1976 book *Electric Wars*, page 180: "Once radioactive materials have decayed past the point at which their internal generation of heat is a dangerous self-contained means of dispersal—in a few decades, or well within a single human lifetime—it is difficult to see in what way they are any more or less hazardous than other poisons produced by industry."

Risks from Alternative Energy Sources

Nuclear power is unique in that safety was considered right from the start, before there were even any power plants. In view of nuclear power's historical origins in weaponry, this is perhaps not surprising. But it is unprecedented. In other fields, the product generally went out to the public, and it wasn't until it became clear there was a hazard that safety became an issue.

When I was young, some of my brother's high school classmates built airplanes and flew them in the races at Curtis Wright Airport near by. No one asked what their qualifications were to build or to fly aircraft, and people were killed in those races, until the races were finally stopped. When my father bought his first car, he was given a few minutes oral instruction by the car salesman, and then turned loose on the streets. No one had made any calculations to show that 50,000 people would someday be killed each year in these machines, and that maybe drivers should be licensed.

But nuclear power engineers started making the kinds of safety evaluations we have been talking about, with detailed technical procedures, resulting in 30-volume safety analysis reports and environmental impact statements of similar bulk. Ironically, this unprecedented effort on the part of nuclear engineers to identify and deal with every kind of anticipated hazard in advance did not create a sense of public confidence. Instead, the reaction seemed to be: *My gosh, they must really be scared of this stuff. It must be terribly dangerous!*

The same kind of calculations have been made for other power sources such as solar or hydroelectric, and there is surprisingly little disagreement about the answers they yield, even between advocates and critics of the various types of power. Take solar power, for example. The most important characteristic of sunlight is that it is a *dilute* energy source. No advance in technology can increase the amount of energy in the sunlight striking each square foot of earth, about ten watts maximum, at noon on a sunny summer day. Thus, to produce as much electricity as a modern power plant, a solar power plant must have collector systems covering about 20 to 50 square miles of ground. If we choose to capture the sun's power through windmills (wind energy comes from air being heated unevenly by the sun), we will need a 200 square mile area containing a thousand towers 200 feet high, just to replace one nuclear or coal-fired plant.

This means a direct solar plant needs 20 times more steel than a nuclear plant, 13 times more concrete, and twice as much construction labor. (Building a large solar electric plant is more expensive than building a nuclear plant to produce the same amount of electricity.) We know how many people get killed, on the average, producing a ton of steel or pouring a cubic yard of concrete. Furthermore, all those mirrors and solar collectors must be kept clean or the plant efficiency drops off, and falls are one of the biggest causes of death.

We have no basis for assuming that people cleaning panels at solar plants will fall any less often than people elsewhere. Construction materials must be transported by truck or train and rigged into place with cranes. All these operations have a known number of deaths per hour or per ton. I'm not defending the reality of these deaths. I'm just saying they're exactly the same sort of calculated deaths claimed for nuclear power. You can argue that nobody has been killed yet at any of the experimental solar units (which may or may not be true—I don't know), but I would respond that no member of the public has been killed by nuclear power.

The bottom line is that calculations like this have been made by persons of various persuasions, and they keep showing that nuclear power is safer than any of its competitors—by a large margin. If you find someone arguing against nuclear power on a safety basis—including waste handling, potential accidents, uranium mining—try to get back to the actual calculation of risks. On this matter there is little disagreement. To be rational, an argument against nuclear power must be based on considerations other than safety or public risk.

Putting a Dollar Value on Human Life

Figure 9.3 shows estimated costs of various life-saving measures ranging from vaccinating children to controlling emissions at uranium-handling facilities. These figures were calculated by taking the cost of the action—the vaccination or the emission control system—and dividing by the number of years of life estimated to be saved. The numbers run from a net gain for a program of expanded immunization, to $34 billion cost of each year of life saved for a uranium handling facility.

Type of Life-Saving Measure	Cost per Year of Life Saved*
Childhood immunization for all (vs. scattered efforts)	less than $0†
Flammability standard for children's sleepwear size 0-6X	less than $0†
Flu vaccinations for all citizens	140
Mandatory motorcycle helmet laws	2,000
Prenatal care for pregnant women	2,100
Chlorination of drinking water	3,100
Supplemental food program for women, infants and children	3,400
National (vs. state and local) 55 mph speed limit	6,600
Improved basic driver training	20,000
HIV/AIDS treatment	20,000
Utility pole spacing decreased to 20 per mile (vs. 40)	64,000
Asbestos banned for pipe insulation	65,000
Air-bags (vs. manual lap-belts) for cars	120,000
Seat belts for school bus passenger	2,800,000
Radon emission control at Department of Energy facilities	5,100,000
Asbestos banned for roof coatings	5,200,000
Chloroform reduction at 33 worst pulp and paper mills	57,000,000
Radioactive emission control at NRC-licensed facilities	2,600,000,000
Emission control at uranium-handling facilities	34,000,000,000

*Since we all die eventually, the idea of "a life saved" is cloudy. Therefore we estimate the number of years that each death foreshortens a life. For example, if average life expectancy is 77 years and the average age of death of an automobile accident victim is 35, each death costs 42 years of life. The very high numbers result from measures, which are believed to save only a small fraction of a year of life.

†The cost of these life-saving measure is actually less, in dollars, than paying for the treatment of children who would otherwise be stricken with disease or fire.

This data from The Center for Risk Analysis of the Harvard School of Public Health, in a 1995 study with 1316 references, funded by the National Science Foundation and others.

Figure 9.3 The Cost of Various Life-Saving Measures

By *net gain,* they mean that it costs fewer dollars to immunize children, for example, than to care for the sick and the dying that would otherwise occur. By *years of life saved,* they estimate the number of years that each accident or illness shortens a life.

Some people react with horror to such a calculation. *How can you put a dollar value on human life?* they ask. I have heard people argue that when calculations show that a power plant may ultimately cause a number of deaths from air pollution, construction and mining accidents, and the like, that approving such a plant is tantamount to cold-blooded murder. That, of course, is nonsense. Building any other kind of power plant will also cause a certain number of deaths, and building no plant and leaving people with an energy shortage can be shown to lead to even more deaths.

I realize that there is a natural human aversion to assigning a dollar value to human life. But we do it all the time. It is unavoidable. When air-bags were first offered as an option on private automobiles, very few people chose to spend their money for this added safety. By this decision, they put a dollar value on their life. Similarly, people buy smaller cars to save money on gasoline, although they increase the chance of killing themselves and their passengers by doing so. So we should not pretend that making such calculations is something no humane person would do.

However, there is an important objection here, which must be conceded: There is a difference between choosing to take a voluntary risk and finding that a risk has been foisted off on you unavoidably. We can choose to ski or scuba dive, we can choose to smoke, drink or get fat, and still legitimately object if we believe we are being endangered involuntarily by the action of others. But in this interdependent world, we can hardly escape being at the mercy of others—airline pilots, bus drivers, and car mechanics, not to mention other drivers, food servers and drugged gunslingers, any of whom may maim or kill us. So long as we depend on central station power plants, we are exposed to whatever risks they pose. It is, therefore, in our interest to understand the relative risks associated with plants of each kind.

There is another way of looking at this question. Suppose the average worker earns $50,000 a year and has a 45-year productive life (from age 20 to age 65, producing goods or services for others). During this period he or she has been paid 45 x $50,000 = $2,250,000 for the productive output of a lifetime. If we decide to add another safety system to a plant at a cost of $2,250,000 (and that doesn't buy much hardware in today's construction market), then we have swallowed up one person's entire lifetime output.

In a real sense, we have judged that the calculated lives saved are worth that much money. In another sense, we have bought one real, living person's life output to trade for a hypothetical life, somewhere in the future. (Some economists argue that the money not only paid for the safety system but also supported grocers, teachers and others who sold goods and services to the system-maker. True. But these same people would have been supported if the system-maker had been making something useful, so I consider that that aspect cancels out, and the basic argument remains valid.)

I hope it is clear that I am not against safety systems. I am just trying to explain that you can't know whether a given system or device will actually increase a plant's safety until you analyze it functioning within the entire plant.

"Lies, Damned Lies, and Statistics"

There is one more tool you're apt to need in talking with scientists or engineers, particularly about safety, and that is statistics. The fact that statistics exist reveals one of the most important secrets about the whole scientific enterprise. It's something scientists don't talk about much, even among themselves. The secret is this: science can tell us nothing whatsoever about any real person or cabbage or electron. Science can only deal with averages or means or typicals. And these are not real entities; they're figments of the imagination, metaphors, and symbols.

There are about 280 million Americans, male and female, and among them they have 280 million breasts and 280 million testicles. Thus the average American has one breast and one testicle. Doesn't sound like most of the people I know, who I thought were pretty average!

Averages

Suppose everybody on your street makes exactly $10,000 a year. (So it's a poor neighborhood; this makes the point easier to see.) There are twenty of these poor families, and they pretty well define the character of the neighborhood. Then a millionaire buys up three lots (buying out three families) and builds a mansion there. Before he moved in, the average income was $10,000, but now it's 10,000 times 17 plus 1,000,000, all divided by 18, which comes out $65,000. Pretty classy neighborhood now, eh? Based on this new prosperity, you might think the neighborhood would no longer be eligible for the free clinic, taxes would go up and Meals on Wheels would no longer visit. A triumph for scientific analysis, but a disaster for the real people there.

Now, scientists are not oblivious to this sort of problem, and that's why you don't see averages quoted very much any more. The *average,* or *mean,* as you know, is calculated by adding up all the incomes and dividing by the number of people, as we just did in the preceding paragraph. In cases like this, scientists are apt to use the *median* instead of the average. The median is the number lying right in the middle of the data (like the median strip on a highway), and is closer to the popular idea of *typical.*

There are as many people earning more than the median as there are earning less. In this case, the median income is $10,000, which certainly represents the typical inhabitant, but it gives no hint at all that the millionaire exists. To avoid such problems in the real world, government statisticians have evolved some very complex mathematical techniques to ensure that pockets of poor people in a generally rich county will still get the public assistance they need. But it can't be calculated simply.

If you really want to understand all the quirks of statistics, there's a lot to learn. People get Ph.D.s in statistics and write long difficult books about it. All I can do here is to point out a few of the pitfalls, so you won't be so easily misled by such things as averages.

Probabilities

First, let's note the difference between *probability* and *odds*. Probability is the absolute chance that something will occur, whereas odds are the ratio of one event versus another. They're a slightly different way of looking at things. Gamblers talk about odds, but scientists and engineers like to work with probabilities.

A hard fact to swallow is that most things that happen are long-shots—highly improbable. The probability of getting any particular bridge hand—*any* hand, not just a winner—in the same order, is one out of 52 x 51 x 50…for 13 numbers, which is the number of cards in a bridge hand. The product of multiplying these 13 numbers together is a very large number—a *huge* number—4×10^{24}. If you're willing to take them in any order, then you can eliminate the six billion hands you can make out of the same cards. That increases your probability of drawing "the same hand" to one in 7×10^{14}. That's one in 700 million-million! You have a much greater chance of getting killed by a meteorite than you have of drawing that particular hand. It is a very improbable event and will probably never happen to you again. But it is not *surprising,* because *any* particular hand is equally improbable, and you are sure to get one of them. Drawing a particular hand would be surprising only if you had predicted ahead of time that you would draw exactly that hand. So while improbable things happen all the time, they are only noteworthy if they have particular meaning.

Of course, it would also be improbable if the only events that befell you were those without meaning. *Some* of life's improbable events are bound to be meaningful, such as drawing a full house in poker. So, if you have an occasional, meaningful low-probability event happen to you, it would be unscientific to assign it earth-shaking significance. During your lifetime, you're bound to draw a few winners and a few losers, and that's just the way it is.

This situation creates another trap for the unwary scientist, which can be illustrated as follows. It's common to make three experimental runs—to gather three data points—before trying to reach even a tentative conclusion. Suppose you do this. You will get three different answers (experiments on real physical systems never come out *exactly* the same).

Let's say you get 4.30, 4.35, and 5.11 inches, or pounds, or whatever you're measuring. The immediate temptation is to find something wrong with the last run: *That darn meter must be acting up again* or *Maybe I didn't shut that last run off as sharply as I did the other two. Yes, I'm sure that's what happened.* So now you feel justified in eliminating the last data point (the 5.11), and you have 4.30 and 4.35. Excellent agreement! And you decide that two points is OK, considering how much work went into getting each one.

The fact is, and it's easy to show statistically, that it is much more likely that three data points will include two numbers close together and a third quite divergent, rather than having them all nicely spaced equidistant, as you might intuitively expect. It's good to use your intuition and common sense, but sometimes they can lead you astray. The proper procedure for the above experiment would use all three numbers, not just two. It's bad science to dismiss an experimental result just because you don't like the answer. A surprising result might cause you to re-examine the whole experiment, and if you find that the run with the divergent answer was different in a way that explains the divergence, then you've learned something. You could run further experiments, taking the new information into account. But adjusting data to bring answers into line is a basic no-no in science.

I'll give one more example, often cited, of how statistics can sometimes give an unexpected answer. Let's assume there is a test for the AIDS virus, HIV; a test that is quite cheap and 98 percent effective. Sounds good, eh? Well, let's see. First we have to find out what is meant by "98 percent effective." For this case, it means that for every 100 tests, 98 will indicate correctly and 2 will be wrong. Suppose also, for the purpose of this example, that one person in 1,000 is actually HIV-positive. Knowing all this, you take the test and are told that you tested positive. What are the chances that you really are?

The answer is not mysterious; you just have to proceed carefully. Suppose we test 100,000 people. We've said we can expect about 100 of them (1 in 1,000) to be HIV positive. The test will show 98 of these positives and 2 (false) negatives. That leaves the 99,900 people who do *not* have HIV. Ninety-eight percent of these (97,902) will correctly test negative, and 2 percent (1,998) will show (falsely) positive. Only 98 out of 98-plus 1,998 positive tests, or 4.68 percent, validly indicate a true HIV positive condition. Thus, *a test properly described as 98 percent effective is, in fact, wrong more than 95 percent of the time.* That's statistics!

(Don't be too hard on yourself if you had problems with this. I had to call on world-class statistician, Jessica Utts at UCal Davis, to bail me out.)

The Constant-Sum Game

If you go on a low-fat, low-sugar diet and get lots of exercise, you may lose weight, but you will thereby increase your chances of being killed in an automobile accident. *How can that be? Am I really more vulnerable when I am slim and trim?* No, of course not. In your newly fit condition, you might even dodge a car that would otherwise have hit you. So your chances of dying by car accident in any given year are as low or lower than before. And, with your healthful lifestyle, your chances of dying of a heart attack are less. So you will probably live longer. But here's the rub: Mortality is what mathematicians call a constant-sum game. Your chances of eventually dying from one cause or another are exactly 100 percent, no matter what you do. Lowering one potential cause of death can thus only raise the chances of something else getting you sooner or later. The reason we choose a particular activity is because we hope it will happen later.

Chernobyl

Sooner or later, we have to talk about Chernobyl. I have given it short shrift here because no one would suggest building more reactors like the one that burned so ferociously in Russia in 1986 and scared the whole world with its radioactive cloud. It was an inherently unstable design, with no leak-tight containment structure around it that could never have been licensed in America, Western Europe or the Pacific Rim. The accident was caused by a casually planned, poorly-controlled experiment, ironically, to get some reactor safety information, and automatic safety devices had been deliberately disabled. The hope was that the plant would be able to ride through such a test and prove itself unusually self-protecting. It didn't. The hope was misplaced.

The only reason that other reactors of similar design (all located within the former U.S.S.R.) have not been shut down is because their countries are in desperate need of the electricity they produce and it would be a severe hardship to deprive them of it. Lithuania, for example, gets more than three-quarters of all its electricity from Chernobyl-type reactors. Reactors of this type, with certain modifications, can be operated safely if strict discipline is maintained. They will eventually be phased out as soon as they can be replaced by American-type nuclear reactors or fossil-fueled plants.

> We should, however, clarify the health impact of that accident. Reports in the media, and even from ostensibly credible sources, have been apocalyptic and filled with wholly false numbers. We can ignore the *NY Post* report of "15,000 human bodies pushed down by bulldozers into the waste pits" and the *National Inquirer* description of a six-foot tall mutant chicken caught in the forest near Chernobyl. But what are we to make of a 1995 Reuters dispatch that 800,000 children have been affected by the consequences of the Chernobyl accident, which was "as terrible as a nuclear attack"? Or the December 2000 Associated Press report that "3.4 million of Ukraine's 50 million people, including 1.3 million children, were exposed to radiation....
>
> According to Ukranian government figures, 70,000 people in Ukraine alone have been disabled by radiation ... 1,040 square miles of the most poisoned land on earth ... if just a speck of radioactive dust were to enter my body, it could kill me." Or the April 2001 *Toronto Star*'s Sunday article that begins, "It is a catastrophe of global proportions: a silent unseen killer is slowly creeping its way out of Belarus and into surrounding countries. It is destroying the future of 10 million people in Belarus who struggle daily with the effects of radiation.... The countryside is poison now....There are no birds in the trees ...And until the world is ready to accept the reality of the situation in Belarus, the

> population will continue to slide toward extinction—and the fallout will spread..."

Where do they get such stories? The answer is simple: from top Ukranian government officials. In 1995, the Ukranian Ministry of Health issued a press release stating that the Chernobyl accident accounted for 125,000 fatalities. What are the facts?

The Facts

A cumulative total of twenty-eight people have died from radiation and two from scalding, all of them firefighters, plant operators or others involved in trying to bring the situation under control at the plant. This figure of 30 deaths comes from the *United Nations Scientific Committee on the Effects of Atomic Radiation* (UNSCEAR) 2000 Report to the U.N. General Assembly. That report states, "Fourteen years after the Chernobyl accident there is no scientific evidence of increased cancer incidence, increased mortality or the occurrence of other diseases attributable to radiation."

Much has been made of the apparent increase in thyroid cancer in children. Fortunately, this is a condition easily treated. Moreover, since there seems to be no correlation between these cases and the amount of radiation exposure the children received, and since there has been no increase in leukemia in the whole exposed population (leukemia being the first symptom of radiation exposure expected), there is reason to believe that something else is at work. A likely cause is that screening of this population for the first time may have revealed normal "hidden nodules" that exist in many people and cause no symptoms.

In addition to impact on human health, there is the question of contamination of the environment. Apocalyptic descriptions of enormous areas of land made forever uninhabitable have been circulated, some from ostensibly credible sources. Let us again turn to the most authoritative source, the UNSCEAR 2000 report. Its maps of radioactivity show that nearly all of the evacuated area poses no threat to human health, being no more radioactive than many areas of the world where people have lived healthily with high natural radiation background. The numbers tell the story:

The *average* radiation level in the contaminated area is only *0.8 rem per year*, with a *maximum* of *3 to 8*. For comparison, note the following *natural* levels:

> World average: *0.24* rem per year; U.S. average: *0.36*
> Parts of the U.S. Capitol and N.Y.'s Grand Central Station: *over 0.5*
> US Rocky Mountain states: *0.6 to 1.2*
> Parts of Sweden: *up to 3.5*
> Guarapari, Brazil: *3.5 to 7.9*
> Tamil Nadu, India: *5.3*
> Southwest France: *8.8*
> Ramsar, Iran: up to *70*

Yet, when the radioactive plume from Chernobyl reached America, the U.S. Environmental Protection Agency announced that the rain over Oregon contained 500–600 picocuries (a picocurie is a millionth of a millionth of a curie) of radioactivity per liter and warned people not to drink rainwater. No one pointed out that beer normally contains 300–400 picocuries per liter, milk contains over 1,200 and salad oil nearly 5,000. A few years previously, the fallout from A-bomb testing in China raised the radioactivity of milk in the U.S. to several thousand picocuries, but there were no health warnings on that occasion. It seems that situations described as harmless when associated with weaponry are viewed with alarm when nuclear reactors are involved.

Science magazine, the journal of record of American science, published the following report in its May 31, 1991 issue, five years after the accident:

> Reports that the Chernobyl nuclear accident caused widespread illness are false, according to radiation experts who gathered here last week for the first comprehensive international evaluation of the consequences of the nuclear explosion ... The independent review, called the International Chernobyl Project, began in 1990 in response to a request for help from the Soviet Union. About 200 scientists and medical experts from 25

> countries focused their research on a group of 825,000 people living in the contaminated areas ... The teams did not find any health disorders that could be directly attributed to radiation exposure ... The analysts found no evidence of a statistically significant increase in fetal and genetic abnormalities, or in leukemia and thyroid cancer ... Radioactivity in drinking water and food was well below levels hazardous to health—in many cases, even below detection limits.

The International Red Cross team of experts sent to assess the situation issued the following words:

> We found a widespread conviction among the population and among the medical profession that there are substantial increases of pathological changes in the thyroid due to radiation exposure.... This belief runs counter to well-established clinical knowledge...Large parts of the affected areas in Byelorussia are iodine-deficient with resultant goiter endemism; thyroid disorders must, accordingly, have been common even before the [Chernobyl] accident. Many doctors seem to support their patients in their suspicions that their symptoms are due to radiation and appear to lack knowledge of scientific facts on matters of radiation protection ... people attribute all their complaints to radiation, clinging to this explanation, which is in line with their worst expectations.

The team of medical experts from the World Health Organization (WHO) echoed these thoughts in their report as follows:

> Scientists who are not well-versed in radiation effects have attributed various biological and health effects to radiation exposure. These changes cannot be attributed to radiation exposure....Attributing these effects to radiation not only increases the psychological pressure in the population and provokes additional stress-related health problems, it also undermines confidence in the competence of the radiation specialists.

UNSCEAR 2000 notes the impact of this situation on the data: "It was concluded that unfavorable psychosocial factors, such as broken social contacts, adaptation difficulties, and relocation, explained the differences between the exposed and the non-exposed groups. No differences could be related to ionizing radiation"

Predictions of large numbers of cancers are based on the fundamentally flawed LNT assumption that any level of radiation, no matter how low, causes cancer. Such calculations "predict," for example, that persons immediately downwind of Chernobyl have an increased risk of fatal cancer of 0.02 percent. Now, the chances of those people dying of cancer without Chernobyl are approximately 20 percent, depending on a lot of things, most of which haven't been measured. So those people must now look forward (if the prognosticators are right) to their chances of getting cancer being increased to 20.02 percent. Now you see why the doctors are saying they will never be able to detect it, even if it's real. But the prognosticators multiply this figure of 0.02 percent by several million people downwind of the plant and get tens of thousands of additional deaths. Of course, as we noted in Chapter 7, most of these numbers of predicted deaths would go to zero if we stopped using the scientifically refuted premise that no amount of radiation is harmless.

The critical factor here is that impoverished Ukraine and Belarus have a strong incentive to blame, as far as possible, all illnesses on the accident. They inherited from the U.S.S.R. an obligation to pay all Chernobyl victims various pensions and social privileges, and over 3 million people have now been judged to have some "permanent health detriment caused by Chernobyl radiation." Like similar handouts promised by former U.S. Energy Secretary Richardson and hurriedly rushed into law, no politician will speak out against such largess. For Belarus alone, this burden is estimated to add up to $86 billion by the year 2015. For health officials sorely strapped for resources to support their people, such a money barrel is too hard to turn away from. Unfortunately, it is also conceivable that some of this money will end up in the otherwise nearly empty pockets of those same, or other, officials.

Hiroshima

A similar game is being played in tabulating up the radiation victims of Hiroshima. On the 1995 anniversary of the bombing, *The*

Washington Post ran a page one story stating: "This year, the names of 5,094 people, mostly victims of radiation-caused leukemia, were added." These figures are apparently based on all of the deaths from all causes in the area. A more realistic figure is 425 cumulative deaths over 50 years for survivors of the blast, from radiation attributable to the bomb, as estimated by the Japanese-American team of physicians and scientists monitoring the survivors. Again, even this figure assumes the notorious linear no-threshold premise. The real number is almost certainly less. In fact, the data show that the irradiated survivors are outliving the unirradiated control population, as hormesis is demonstrated once again.

Nuclear Terrorism

Since September 11, 2001, every facet of our activities is being examined as a possible target of terrorist attack. If a terrorist could somehow get hold of a nuclear bomb and detonate it here, the consequences could unquestionably be serious. But that is outside my discussion of nuclear power. The "dirty bomb" scenario, where a conventional explosive is wrapped with radioactive material, is a possibility, but it can easily be shown to be a very ineffective weapon. A significant amount of radioactivity would require tons of shielding just to carry to weapon to the point of detonation. Scattering the material with explosives would not cause a serious public health problem unless the material could be finely dispersed. This is not a simple matter. It is unlikely that a dirty bomb could cause a significant number of casualties. There would be no reason to panic unless people have been repeatedly told that there would probably be widespread panic. This has been happening, but we could stop it.

An attack or an accident with a spent fuel shipping cask has been widely heralded as a problem. But these casks are virtually indestructible, having been tested with explosives, fire, impact and water. In addition, they are massive, requiring special trucks to move them; they are not easy to run off with. And like the dirty bomb, there can be no hazard unless some sophisticated method of dispersement is developed, and even then, the effects would be limited.

Many fantastic scenarios have been suggested whereby a terrorist, by attacking a nuclear plant, could allegedly *kill tens of thousands of people* and *render vast areas of land uninhabitable for centuries*. Let us skip the particulars and assume that a meltdown has been caused—that's the worst case. As we have discussed, even the Chernobyl reactor, going supercritical and exploding and burning in mid-air, could not cause any public deaths. For an American reactor, the Three Mile Island accident is a more appropriate model. Even if you assumed the containment was breached, analysis of that accident shows that nearly all the biologically significant radioactivity would remain dissolved in the water and plated out on the inside of the plant structure. There seems to be no credible way that a serious public hazard could be created.

Some Final Thoughts

Risk is unavoidable. All we can do is decide for ourselves that any given risk is acceptably low, compared with other risks we face (such as becoming overweight, drinking coffee or not fastening our seat-belts). Calculation of risks posed by various activities has become a standardized procedure. These calculations show that all the major forms of energy production pose less risk than many other activities we willingly accept.

There is fairly good agreement as to the comparative risks posed by different types of power plants, taking into account the whole chain of events associated with each plant, from mining and transporting fuel, plant operation, possibility of accident and handling waste products. These calculations show that nuclear plants pose less of a risk to the public and to the environment than other forms of energy production. Risk analysis enables anyone to compare and evaluate facts based on real world experience, rather than having to rely on speculation and vaguely based fears. One can argue about any of the numbers used, but at least you are then arguing about facts that can be evaluated by any intelligent person.

TODAY

TONIGHT

TOMORROW

TV: Page 23

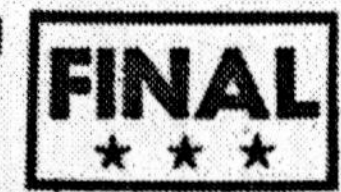

SATURDAY, MARCH 31, 1979 25 CENTS

RACE WITH NUCLEAR DISASTER

Baffled scientists struggle to ward off A-plant meltdown

Figure 10.1 The Disaster That Hurt No One

10. Learning from Three Mile Island

The Humbling Experience

One of the most important lessons I learned from my Naval Reactors experience surprised me at the time, though in retrospect it seems more and more reasonable. We called it "the humbling experience." Before a submarine crew took a ship out for the first time, they had acquired a great deal of operating experience. First, they had all graduated from submarine school, which in itself involved firefighting, battling torrents of high-pressure water from simulated piping failures, being rescued from under fifty feet of water, and other spine-tingling casualty situations. Then, after six months of grueling academic studies, they were required to qualify on all appropriate watch-stations at a submarine prototype propulsion plant. If they were going to a new ship, they would arrive at the shipyard and work through several months—sometimes a year or more—checking as the final construction and equipment installation was completed. Then the test program would begin: cold system tests, hot tests, physics check-outs and the slow buildup to full power—all alongside the dock.

When everything finally appeared to be satisfactory, the last step before sea trials was the "fast cruise." All pipes and wires to shore facilities were disconnected except for the heavy ropes that held the ship fast to the dock, and a two-week simulated cruise was run to put the plant through her final paces. By this time, it was a rare crew indeed that had not had at least one real goof—an embarrassing experience that left the crew's confidence severely shaken. But once in a blue moon, a crew proudly announced they were ready for sea trials and they had not had a single goof. Their performance over the demanding tests of the past many months had been flawless.

I went out on all initial sea trials (except for a few where I had schedule conflicts) and you would think that this flawless crew would be one I looked forward to going out with. But in fact, I learned that a crew that had never once had a humbling experience was a scary one. They were too cocky, too sure of themselves, and their first stumble was apt to be a beaut. All our operators were good; they were in the top few percent of their class and had proved themselves through the nuclear power curriculum. They also knew that the other crews were good and that the other crews had blown it at least once. When they themselves reached sea trials without a blunder, they were busting their vests. I preferred to go out with a crew that knew they could go wrong and were determined not to let it happen again. T. H. Huxley said it well in his book *On Medical Education:* "There is the greatest practical benefit in making a few mistakes early in life."

A similar situation developed in the commercial nuclear power program. The industry had taken extraordinary measures to cover every contingency, and by early 1979, there had never been a casualty leading to a public hazard. (Never mind "The Day We Almost Lost Detroit" and the hyped-up Browns Ferry fire. There was absolutely no release of radioactivity from the plant in either case. Those incidents never came close to endangering the public.) Seventy-two nuclear power plants had run for a cumulative total of 400 years in the United States, and another 92 had passed their safety reviews and were under construction. Twice that many naval reactors and another 200 commercial nuclear power plants elsewhere in the world were performing safely. Everyone was feeling pretty smug. But there is no basis for the oft-repeated charge, "This was the accident they told us could never happen."

The widely discussed official NRC reactor safety study "An Assessment of Accident Risks in U.S. Commercial Nuclear Power Plants (1975)" suggested publicly that an accident of this magnitude was about due.

The stage was set for a humbling experience.

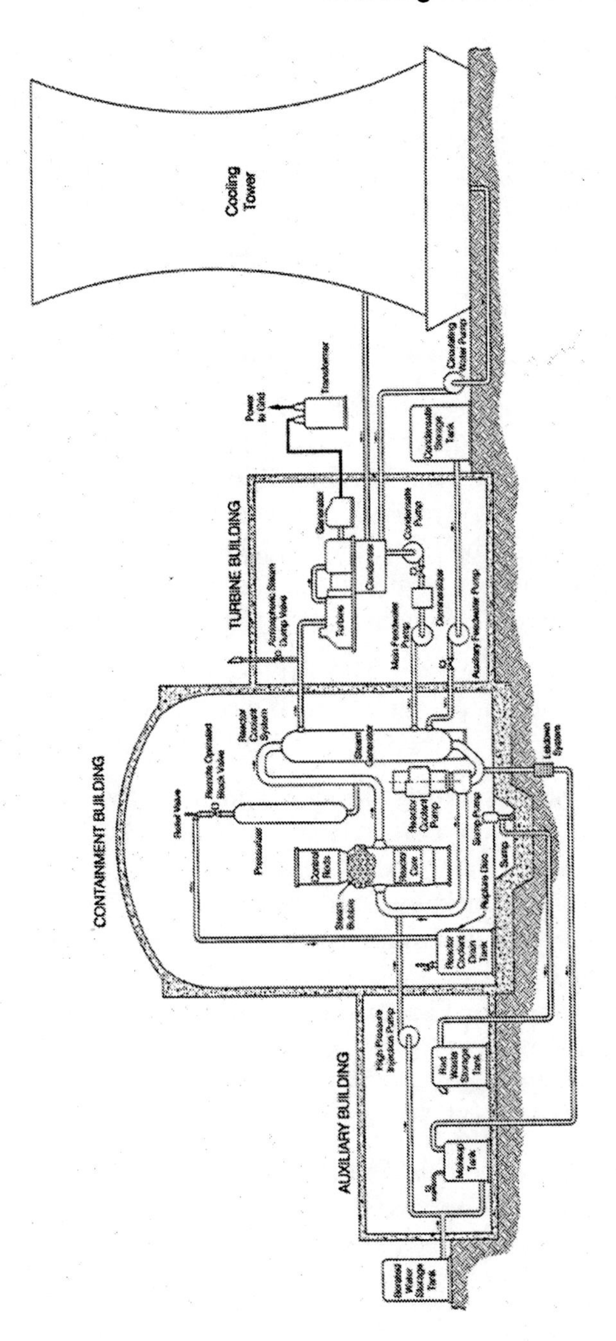

Figure 10.2 Layout of TMI-2 Plant (*DOE*)

"The Island"

There is nothing nuclear or atomic-looking about the Three Mile Island plant. There are no eerie glass cylinders pulsing with glowing purple warp-drive engines; just a lot of white-painted insulation covering piping, valves, pumps and pressure vessels of various sizes. The pumps make a steady hum, the turbine makes a louder hum and the low roar of the ventilation fans is everywhere. But it is all very ordinary. If you don't know what to look for, you could easily take it for a chemical plant or an oil refinery, except that there are no unusual smells.

It is ironic that the word *plutonium* links atomic power with visions of hell in many people's minds. That image more accurately belongs to the old-fashioned coal-burning power plants (many of which are still operating), with their roaring furnaces and angry flames boiling out of burned-out places in the flues. In the old coal-burning plant, there is no place to get away from the heat and the soot, and workers finish their shifts with torrents of sweat caking the grime to their skin and clothing.

The TMI plant is spotless. The control room is like all modern control rooms, air-conditioned, quiet, with bright, uniform fluorescent lighting. It could belong to any modern factory or laboratory.

In nuclear plants, as in long distance airplane flights, a big problem is how to keep the operators alert, when the plants run along for hours with little need for human intervention. This problem ended suddenly at the TMI-2 plant, early one morning in 1979. (Why do such things always seem to happen in the pre-dawn morning?) TMI-2 is the second plant at Metropolitan Edison's Three Mile Island Nuclear Power Station, in Londonderry Township, near Harrisburg, Pennsylvania. The sister plant next door, TMI-1, had been reliably delivering electricity to the grid for four and a half uneventful years, and the two plants were being counted on to supply the electrical needs of 300,000 homes.

That morning, TMI-1 was getting ready to start up again after being shut down for several weeks for normal refueling. To give you a feeling for what it was like to be there when a nuclear reactor melts down, I will describe some of the events of the first two minutes of the accident, taken from the operating logs and interviews with the participants from the many investigations that followed.

Figure 10.3 TMI Control Room

MARCH 28, 1979, 4:00 A.M., CONTROL ROOM OF TMI-2. Ed Frederick and Craig Faust, control room operators at TMI-2, are performing routine functions, "charging up water to compensate for some leakage we've been having through the relief valves," explains Frederick. At this time of night, there are no messages blaring out from the paging system loudspeakers. Bill Zewe, shift supervisor at TMI-2, looks out of his office into the quiet control room, pours another cup of black coffee and mutters to himself: "They sure knew what they were doing when they called this the 'graveyard shift.'" He has no way of knowing that in the next few minutes more than 100 of the control panel's 750 alarms would go off, signaling the beginning of a casualty that would put TMI in screaming headlines around the world as the site of the "world's worst nuclear power plant disaster."

TMI-2 started up just three months previously, and is now operating quietly in automatic between 97 and 98 percent of full rated power. There are a few red warning lights lit on the panel, but the operators understand that these have been on for some days because of some minor instrument problems, and they are causing no trouble.

Fred Scheimann, the Shift Foreman, is at the condensate polisher in the auxiliary building, helping to unclog the resin. The condensate polisher is a device like an ordinary home water-softener that uses a special fine granular resin to remove small quantities of impurities from the water. After a period of use, the resin granules become rather hard-packed and have to be fluffed up by bubbling compressed air through them.

He doesn't find out until much later that this time the air pushed some water past a small check valve in the pneumatic instrument tubing, causing the inlet and outlet valves to the polisher to close. This in turn interrupts the flow of make-up feedwater to the reactor, and causes condensate pump 1A and its booster pump to automatically trip off the line. After one second, this causes both of the main feedwater pumps that supply water to the steam generators to shut down. The main turbine that runs the electric generator that creates the electricity for which this plant was built is suddenly deprived of steam and automatically shuts down. One second after that, the three emergency feedwater pumps (one steam-driven and two electric-driven) start up automatically.

4:00:36 AM, SHIFT SUPERVISOR'S OFFICE. Bill Zewe hears the annunciator alarm on panel 17, followed a second later by the alarm on panel 15 and then the alarm on panel 5. Hurrying into the control room, he sees the status lights on panel 5 telling him that condensate pump CO-P1A and feedwater pumps FW-P1A and P1B have tripped, shutting down the main turbine. The status light on panel 4 tells him that all three emergency feedwater pumps have come on, but he does not realize that block valves EF-V12A&B are closed, preventing any water from entering the system. (No one is ever able to explain who shut these valves, or why. They are normally open. But at this point there is still one to one and a half minutes before the steam generator boils dry.) The alarm printer that prints out details as to plant events and status begins receiving over 100 alarms per minute and soon gets so far behind that it is of no use to the operators. Craig Faust turns to his fellow-operator Ed Frederick and says. "We're in trouble—something's going wrong in the plant."

4:00:40 A.M. With the turbine no longer removing heat, the reactor system pressure rises and the status light on panel 4 comes on, indicating that the Pilot-Operated Relief Valve (PORV) has opened to relieve the buildup of pressure. The valve relieves to the coolant drain tank, which begins to build up pressure and temperature, as expected.

4:00:45 A.M. The annunciator on panel 8 goes off, and the status light and meter on panel 14 show that the reactor pressure has risen another 100 pounds, which shuts down the reactor on a high-pressure signal. (The reactor is now shut down, but decay heat from short-lived fission products continues to give off 160 MW (million watts), about 6 percent of full reactor power. This decay heat will drop to 33 MW after one hour and to about 15 MW at ten hours. It drops more and more slowly after that.)

4:00:50 A.M. With the reactor shut down, the steam generator cools the reactor water and the water begins to shrink. Faust shuts off the valves that had been draining off water for routine continuous purification and turns on make-up pump MU-P1A. He does not notice that the status light does not come on to indicate that the pump started; in fact, the pump did not start. (Later tests indicate that the pump circuitry is apparently normal. Faust may just have not held the switch handle over far enough or long enough.)

Reactor system pressure has now come down to the point where the PORV relief valve should close. The status light goes off, indicating that the electric current to the solenoid that opens the valve is now off. But there is no indicator light on the panel to tell whether the valve has actually shut. It has not, and water continues to empty out of the reactor system and into the drain tank at the rate of 220 gallons per minute.

Figure 10.4 Herblock's suggestion for TMI control room modification. from *Herblock on all Fronts* (New American Library, 1980)

This is the first crucial error in the accident chain. It is the prime cause of the reactor coolant draining down to where the core was no longer covered with water and began to melt from the decay heat being generated. The usual means of checking that the PORV is closed was to see that the temperature at the valve outlet returned to normal. But in this instance, the valve had been leaking, and the operators assumed that the high temperature they continued to read must have been due to this leakage.

4:01:07 AM. The high temperature alarm on the PORV outlet sounds, but this is assumed to be due to the effects of leakage, added to the recent opening. The reactor coolant system now reaches the low pressure trip point, but this is shown only on the strip chart and not by any alarm.

4:01:18 AM. Frederick notices that the make-up pump did not start; so he starts it. He opens valves MU-V16B and DH-5A to admit more water. Seven seconds later, the water level in the pressurizer reaches an acceptable minimum and continues to rise.

4:01:37 AM. One minute into the accident scenario, 14 alarms have already sounded and numerous indicator lights have come on or shut off. The alarm printer, telling what alarms have sounded and why, is now running nearly one minute behind. An instrument air line to the hotwell level controller is broken (unknown to the operators), and this and other steam system problems keep distracting the operators from the crucial reactor system problems.

4:02:39 A.M. Reactor system pressure drops to 1,600 pounds, which automatically starts the engineered safeguards system—two high-pressure injection pumps that pour large quantities of water into the reactor system. By now, a wide variety of alarms are sounding, causing everyone's heart to pound and their blood to race as each new sound adds its urgent scream to the cacophony. Alarms are designed to be loud and shrill enough to irritate and ensure they won't be ignored. But they certainly wreck any efforts at calm deliberation. Faust said later, "I would have like to have thrown away the alarm panel. It wasn't giving us any useful information."

The record goes on this way, for a total of 483 entries. What it shows is the mundane ordinariness of the situation. Just valves and pumps and big steel pressure vessels, and instrument dials indicating temperature, pressure and gallons per minute. Of course, unless you are familiar with the layout of the plant and the numbering system for all the valves and pumps, the welter of names and numbers is confusing. It sounds like a jumble of nomenclature, lingo, and high-class slang. But there is nothing mysterious going on; nothing that is too complicated for the average citizen to understand if he or she is willing to make the effort.

The temperature and pressure situation needs a little explaining. Every power plant has an energy source, such as a furnace burning coal, oil, or gas, or a nuclear reactor splitting atoms and heating up water. Every plant also has a "heat sink" (such as a turbine) to carry this heat away and turn it into electricity. As long as the heat is being carried away as fast as it is being created, the temperatures all remain steady. But if either the heat source or the heat sink is impaired, temperatures can change fast. At TMI, the main turbine tripped off the line. Suddenly, the operators had a reactor continuing to produce heat, but no turbine taking the heat out of the steam to generate electricity, so the plant started to heat up quickly. When the plant heats up, the pressure in the cooling water system also rises, and that's just what happened. It quickly reached the pressure where the over-pressure protector shut down the reactor. Still, there was no problem, so far.

Both the reactor system and the steam system have some *thermal inertia*; i.e., after shutdown, the fission products in the reactor core keep giving off some heat for a while. The colder water in the steam generator wants to draw heat from the reactor system until the two fluids reach the same temperature. This brings us to the next point: water at these high temperatures expands quite a bit—up to 40 percent compared with room-temperature water. This extra volume of water has to go somewhere when the plant heats up, and return from somewhere when it cools down. So the reactor cooling system has a separate surge tank hooked up to it called a *pressurizer* that accommodates the expanding water on heat-up and supplies the needed water to make up for shrinkage when the plant cools down.

One other point to clear up is that of relief valves. All closed systems that might in any way become over-pressurized have relief valves. These are simply valves that normally stay closed, but if necessary will pop open and relieve excessive pressure. Your hot-water heater has one, as does your car radiator. These are required by engineering codes, usually backed up by regulations or law or both, in every modern industrialized nation. Because these relief valves must be hair-triggered and never fail to open, they do not always close again tightly after being open. To protect against this problem, TMI, like many nuclear power plants, had additional discretionary relief valves set at a lower pressure, so that in most cases, a pressure surge could be relieved by these valves before a relief valve pops.

In TMI, one of these valves is called the PORV, for pilot-operated relief valve. But what if one of *those* valves failed to close properly? That was easy: if one of those valves failed to close fully after opening, a stop valve in the line could be closed so that the leak could be stopped. Regulations prohibit having a stop valve that could prevent the main relief valves from functioning, but a stop valve in another branch of piping, to stop a non-required relief valve from leaking, is permissible.

"Don't Let 'er Go Solid"

Getting back to TMI on that grim and confused morning in 1979, at two minutes into the accident the reactor system pressure had dropped so low that the high-pressure injection pumps had automatically started and were pouring water into the reactor plant. Unbeknownst to the operators, the steam generators had now run dry because the valves at the emergency feedwater pumps were closed. This caused the reactor water to heat up and expand, causing the water level in the pressurizer surge tank to rise, misleading the operators into believing that the reactor system had plenty of water. Therefore, Frederick shut down one of the injection pumps and sharply reduced the flow from the other. He was afraid that continued input of water would lead to "a solid system." And that requires another explanation.

A *solid system* refers to a condition where there is no steam bubble in the reactor system; it is filled solid with water. Normally there is always a steam bubble in the pressurizer, created by a large electric heater put there for that purpose. This heater has two functions: first, to provide a cushion of steam to accommodate surges of water into and out of the system when it expands or contracts. (If the system were filled "solid," such surges would cause huge pressure changes that could damage the plant.) The second function of the heater is to make the water in the pressurizer hotter than the rest of the water, so that only the pressurizer water will boil; the rest of the system stays liquid, which ensures adequate cooling for the reactor.

But about five and a half minutes into the accident, the overheated, under-pressurized reactor began to boil and form a steam bubble in the reactor vessel, pushing more water into the pressurizer. The operators now had a wrong picture of the situation on several counts. There is no meter or dial to tell them where the water level is in the reactor vessel, since it is always supposed to be full. The rising level of water in the pressurizer seemed to tell them there was too much water coming in, and they opened a valve to the let-down system to drain some off. At about eight minutes, Faust discovered that the valves to the feedwater system were inexplicably closed, and he opened them, allowing water to rush into the steam generator.

At 4:11 A.M., eleven minutes into the accident, an alarm signaled high water level in the sump of the unmanned containment building, an indication that water was leaking onto the floor from one of the systems. At 4:16, the rupture disk (an overpressure relief device) on the drain tank receiving leakage from the PORV blew open, releasing water and gases into the containment building. Pressure and temperature in the containment rose, as did the radiation level.

"Shut Off the Coolant Pumps"

Shortly after 5:00 A.M., the reactor coolant pumps started to vibrate badly, and at 5:14 the operators shut off two of the four pumps. About 5:30 they shut off the other two. The operators did not realize that the vibration was caused by steam entering the pumps, another indication that there was boiling in the reactor. The reactor fuel was no longer covered with water, and the cooling flow had been stopped. By 6:00 some of the fuel rods overheated and ruptured their zirconium cladding, releasing fission product gases into the reactor water that was still pouring out of the PORV onto the containment floor. Radiation alarms went off in the containment building.

At 6:22 A.M., closed the PORV block valve and stopped the flow of water from the reactor system, but for almost another hour no one started up the injection pumps to replace the lost water. The top of the reactor had been uncovered (and thus poorly cooled) since about 6:15, and radiation levels in the reactor containment were rapidly increasing. The containment building is always closed off and unmanned during operation, but high radiation levels in the containment air present a potential for leakage of radioactivity to the environment. By 6:48 there were indications that as much as two-thirds of the reactor core was uncovered.

Calculations showed that parts of the core may have reached 4,000°F, and examination of the reactor years later showed that several tons of fuel elements melted and dropped to the bottom of the vessel. At 6:45 one of the reactor coolant pumps was turned on, but it vibrated so badly that it was turned off 19 minutes later.

TMI Calls a Site Emergency

At 7:00 A.M. Gary Miller, TMI station manager, arrived on site. A site emergency had been called, which is required by the TMI emergency plan if a situation arises that threatens "an uncontrolled release of radioactivity to the immediate environment." Miller had had several telephone discussions with people at the site, and when he arrived he assumed command as emergency director. He formed teams to cover various aspects of the situation:

- operator activities in the control room;
- review of all procedures to ensure compliance;
- monitoring of radiation on-site and off;
- technical support and communication;

- emergency maintenance; and
- liaison with the utility's home office.

In accordance with the TMI emergency plan, an emergency control station was set up in the control room of the sister plant, TMI-1.

Outside Officials Are Brought In

As prescribed in the TMI emergency plan, Pennsylvania state officials were notified, as was the U.S. Department of Energy's Radiological Assistance office at the Brookhaven National Laboratory on Long Island. But the U.S. Nuclear Regulatory Commission was harder to reach. When the NRC regional office at King of Prussia, Pennsylvania, opened at 7:45 A.M., Miller had already escalated the site emergency to a general emergency, "an incident which has the potential for serious radiological consequences to the health and safety of the general public."

Since the NRC headquarters offices were spread around in eight different buildings in Washington, DC, and near-by Maryland, run by a commission composed of five commissioners of equal authority and unequal expertise, who are prohibited by law from meeting together without previously announcing a public meeting, a potential for confusion and miscommunication existed, which was soon richly manifested. The state sent two radiological monitoring teams, which reported readings outside the plant of less than 1 mrem per hour, well below NRC danger levels.

About 8:00 A.M., the valves closed that automatically isolate the atmosphere in the containment from the outside when pressure begins to build up. Although the whole containment structure is required to meet extremely tight leakage specifications and is retested periodically against those specifications, there are pipes that run through the containment wall into the auxiliary building. Even after isolation, operators felt they had to pump out some of the radioactive water that was accumulating in the containment, and this water went into tanks in the auxiliary building. From there small amounts of radioactive gas escaped into the atmosphere.

At 8:26, the high-pressure injection pumps were turned back on, but even with the high rate of flow, the core was not fully covered again until about 10:30.

The Media Enter the Scene

"Captain Dave," a traffic reporter for WKBO in Harrisburg, tuned in his CB car radio and heard police and firefighters mobilizing in Middletown. He reported this to his station. Mike Pintek, WKBO's news director, called TMI and asked for a public relations official. He was connected to the control room, where he was told, "I can't talk now; we've got a problem."

Pintek then called the utility's headquarters in Reading, Pennsylvania and was told there was a general emergency.

> "What the hell is that?" asked Pintek.
>
> "It's a red-tape type of thing the NRC requires under certain conditions."
>
> "What conditions?"
>
> "There was a problem with a feedwater pump. The plant is shut down. We're working on it. There is no danger off-site. No danger to the general public."

This story, without amplification, went on WKBO's 8:25 newscast. Associated Press picked it up and filed a story at 9:06—six sentences, four paragraphs. Many officials, supposedly in the emergency command chain, heard about the story from the media. Paul Doutrich, Mayor of Harrisburg, got a telephone call at 9:15 from a radio station in Boston. "They asked me what we were doing about a nuclear emergency," he said. "My response was, 'What nuclear emergency?' They said, 'Well, at Three Mile Island.' I said, 'I know nothing about it. We have a nuclear plant there, but I know nothing about a problem'. So they told me; a Boston radio station had to tell me about it."

At 9:15 the White House was notified. At 10:05 the first NRC representatives from the regional office arrived on the site. The radioactivity in the control room air was now above the recommended level for continuous exposure, and personnel there were issued protective face masks with filters. This made communication difficult, leading to further confusion. At 11:00 all nonessential personnel were ordered off the Island. Public statements were now going out from various sources, some asserting that there was airborne radioactivity near the site, others denying it.

Since the operators were unable to run the pumps because of the excessive vibration, they tried to establish natural circulation to cool the core. But this was unsuccessful because the steam bubble at the top of the reactor vessel blocked the flow. The overheated fuel had interacted chemically with the water to form zirconium oxide and hydrogen, and this hydrogen was also part of the bubble. At 11:38 the operators vented some of the gas into the containment building, and at 11:50 a sharp mechanical thud was heard in the control room. It was dismissed at the time as the slamming of a ventilation damper, but it was found later to be a small hydrogen explosion. Workers entering the auxiliary building at noon found very high radiation levels and left after a short time.

By that evening, TMI was on everyone's news broadcast. Walter Cronkite started his CBS Evening News with "It was the first step in a nuclear nightmare ... probably the worst nuclear accident to date."

By the next morning, things looked better. The logbook of the Dauphin County Office of Emergency Preparedness had some reassuring entries:

> 5:45 A.M. Called Pennsylvania Emergency Management Agency—reactor remains under control, more stable than yesterday.
>
> 7:55 ... no danger to public...
>
> 11:25 ... situation same...
>
> 3:30 P.M. ... situation is improving...
>
> 6:12 ... continues to improve, slow rate, off-site release controlled.
>
> 7:00-9:00 ... Island getting better

> 9:55 ... no real measurable reading off-site—no health risk off-site, no emergency, bringing reactor to cold shutdown.

The U.S. Food and Drug Administration began monitoring food, milk and water in the area. Utility and NRC officials began briefing congressmen and others. But a new problem began to develop. It is not agreed as to who started it, but talk of possible evacuation of the area began to be heard. The idea was discussed by the Pennsylvania Emergency Management Agency (PEMA) and shelved, as was the suggestion that only pregnant women and children under two be evacuated.

Soon after the accident, the utility had stopped discharge of water from the laundry facilities, showers and toilets and from leakage from the steam system. This water is not normally radioactive and is discharged directly into the Susquehanna River, but the utility was concerned that it might be slightly contaminated by the accident. However, 400,000 gallons of such water had now accumulated and tanks were close to overflowing. The radioactivity of the water was well within limits, and the local NRC representatives approved resuming dumping. However, as soon as Joseph Hendrie, Chairman of the NRC, heard about it, he ordered the dumping stopped. He was concerned that there would be a public outcry, no matter how little radioactivity was involved. Ultimately, after hours of discussion, he and the Governor's office agreed "reluctantly" to resume dumping, so long as NRC limits were not exceeded.

Conflicting Evacuation Advice

Friday, March 30, the second day after the accident, was a day of considerable uproar, indecision and policy reversals. Although the reactor core was no longer an urgent problem, there was still a large volume of radioactive water and gases to be dealt with. At 7:10 A.M., James Floyd, TMI-2's supervisor of operations, decided to transfer some radioactive gases from the make-up water tank to the waste gas decay tank. He knew this would result in some leakage of gas into the auxiliary building air, and from there to the outside air, but he believed the transfer was necessary. He asked a helicopter to take some radiation readings near the vent stack.

At that moment, Lake Barrett, a section leader at NRC headquarters in Washington, was trying to calculate how much radiation would be received off-site if the waste tank relief valve opened. With the scant amount of information available, he made a stab at it and came up with a number: 1,200 millirem per hour on the ground. Within "maybe 10 or 12 seconds" after announcing this number, he got a call from the site giving him the radiation level measured by the helicopter right near the stack. By a "horrible coincidence," as phrased by the Chairman of the President's Commission on the TMI Accident, the number was also 1,200 millirem per hour. The NRC proceeded without confirming the measurement, without knowing whether it was taken on-site or off, at the ground or by helicopter. In fact, the radiation did not even come from the waste gas decay tanks. The stage had been set for "a morning of confusion and contradictory evacuation recommendations."

After getting the helicopter reading, Harold Denton of the NRC asked his associate Harold Collins to notify the Governor that the NRC recommended an evacuation. Collins did so, adding on his own that the evacuation should cover ten miles downwind of the site. Thomas Gerusky, the state's director of the Bureau of Radiation Protection, called an NRC official at the plant, and they decided an evacuation was unnecessary. Gerusky tried to phone the Governor, but his line was busy, so he hurried over to the Governor's office to talk him out of the evacuation.

At 9:25 A.M., Oran Henderson, director of PEMA, called Kevin Molloy, director of emergency preparedness for Dauphin County, telling him to expect an official evacuation notice in five minutes. Molloy called all fire departments within ten miles and broadcast a warning over radio station WHP. Back at the Island, the utility people heard the broadcast and asked Charles Gallina, the NRC representative, "What the hell are you fellows doing?" Gallina checked with the NRC reactor inspector at the site, who told him, "Things are getting better," and called NRC headquarters "to call back that evacuation notice."

Understandably confused, Governor Thornburgh called Chairman Hendrie of the NRC and was assured that no evacuation was necessary. However, the Chairman suggested off the cuff that the Governor might wish to urge everyone within five miles downwind "to stay indoors for the next half-hour." Later that morning Governor Thornburgh issued an advisory that everyone with ten miles of the plant should stay indoors. At 11:40, Thornburgh received a call from Hendrie apologizing for the NRC staff error in recommending evacuation. Thornburgh asked Hendrie about the suggestion that pregnant women and toddlers be evacuated. Hendrie's reply, played over the speakerphone was, "If my wife were pregnant and I had small children in the area, I would get them out because we don't know what is going to happen." After the call, Thornburgh issued an advisory recommending pregnant women and preschool children leave the region within a 5-mile radius and that all schools within that area be closed.

Through Friday, Saturday and Sunday, the various preparedness offices were in turmoil. The original plans had been for a 5-mile radius, but later they were told to prepare for a 10-mile and then later, a 20-mile radius. The implications of these three plans were quite different. The 5-mile circle included 25,000 people and no hospitals or other facilities that might be difficult to evacuate. The 10-mile circle included several hospitals. The 20-mile circle involved six counties, 650,000 people, 13 hospitals and a prison. Many of these people were Amish, with no telephones, radios or televisions.

The Potassium Iodide Caper and the Explosion Hazard that Wasn't

Meanwhile, the U.S. Department of Health, Education and Welfare (HEW) was becoming concerned about the possibility of airborne radioactive iodine becoming a hazard. It was suggested that if people drank enough potassium iodide solution, this might saturate the thyroid so that it could not absorb the radioactive iodine, and this would protect them. However, there was no source of potassium iodide of sufficient purity for medical use. Saturday morning at 3:00 A.M., Mallinckrodt Chemical Company agreed to supply HEW a quarter of a million one-ounce bottles of the chemical. Working with Parke-Davis in Detroit and a bottle-dropper manufacturer in New Jersey, they got the first shipment to Harrisburg within 24 hours and the final shipment three days later. After a lot of arguing, it was agreed that no one was sure just what medical effects might result from a saturating dose of potassium iodide (even without radioactivity); since there did not seem to be a measurable amount of iodine in the air, the whole subject was quietly dropped.

The final pratfall in this tragicomedy of errors came on Saturday. The NRC people began thinking about the 1,000-cubic-foot gas bubble in the reactor vessel and realized that it was largely hydrogen and that hydrogen was explosive in the presence of oxygen. Calculations showed that if it were to explode, the extent of damage would be enormous, raising fears that resonated with the phrase "could devastate an area the size of Pennsylvania," from the Jane Fonda movie *The China Syndrome*, which was drawing big crowds at the time. Throughout the weekend, this fear was amplified by interviews with almost everyone, from NRC officials and staffers to local high-school teachers, all of whom agreed that hydrogen could make a hell of a bang.

Months later, the President's blue-ribbon panel noted bluntly in its special report: "That it was a groundless fear, an unfortunate error, never penetrated the public consciousness afterward, partly because the NRC made no effort to inform the public that it had erred … the NRC could have determined from the information available at that time that no excess oxygen was being generated and there was no real danger of explosion." In fact, pressurized water reactors, such as TMI, generally keep an excess of hydrogen in the water routinely to suppress any oxygen, because this keeps corrosion of the system to a minimum. The public had been panicked needlessly.

Normalcy returned slowly to the area. Governor Thornburgh repeatedly asked the NRC if the pregnant women and preschool children could return home, but according to the report to the President, "the NRC wanted some specific event as a symbol to announce that the crisis had ended." At first they wanted cold shutdown to be that event, but when it became clear that this would be days away, the NRC finally agreed with the state people that the advisory could be lifted. On Saturday, April 7, the evacuation shelter was closed, and two days later the advisory was withdrawn.

In addition to the Nuclear Regulatory Commission, 23 different federal agencies had gotten involved, from the Federal Preparedness Agency, the Federal Disaster Assistance Administration and the Defense Civil Preparedness Agency, to the National Oceanic and Atmospheric Administration, the Postal Service, the Small Business Administration and the Department of Transportation. In addition, 14 types of state agencies were involved, including one or more emergency organizations from each of 26 surrounding counties (counted as one of the 14 types of state agencies), the Pennsylvania Bureau of Radiation Protection, the Pennsylvania Emergency Management Agency, down through the Bureau of Forestry, the Department of Justice, the Insurance Department, and the Turnpike Commission.

The Seriousness of the Accident

The accident, its causes and its consequences, were investigated in detail by almost everyone. The NRC and its contractors made several detailed investigations, while a special outside panel prepared a 1,500-page report to the NRC and the general public. In addition, President Carter appointed a Special Commission on the TMI Accident that collected reports filling 300 feet of shelf-space in connection with preparing its report. The Department of Energy and the Congress ran their own investigations, and a number of publications, from the Institute of Electrical and Electronic Engineers' professional journal *Spectrum,* to various local and national newspapers and magazines, did their own interviewing and reporting. And numerous organizations with various sociological and political agendas went house-to-house surveying the people. On most of the important findings and recommendations there was a surprising amount of agreement:

1. There was no detectable radiation effect on the physical health of the population or the environment, nor is any delayed effect anticipated. The average radiation dose to the people in the vicinity was a total of 1.4 millirems, which is less than 1 percent of the radiation they get normally each year from natural radiation. By comparison, note that a person who moves from a frame house to a brick house or a stone house

increases his radiation by ten times that amount each year because of the natural radiation of the brick or stone; we rightly assume that the added radiation from such a move is harmless.

2. The highest radiation dose to any of the *plant workers* was approximately 4 rems (4,000 millirems). This is more than the 3 rems per quarter-year routinely permitted workers under NRC guidelines, but less than the permissible 5 rems per year. No detectable health effects were seen.

3. The most serious health threat from the accident was the *mental stress* created for the nearby residents. The on-again, off-again evacuation plans with particular reference to the emotional subjects of fetuses and small children were particularly frightening at a primal level. The continuing pressure of interrogation by survey-takers and media people, asking in effect: *Are you feeling sick yet? Are your children still OK?* was challenging to the sturdiest psyche. Not surprisingly, these studies showed 14 percent of the people in the area drinking more, 32 percent smoking more, 88 percent using more tranquilizers, 113 percent using more sleeping pills in the period right after the accident.

4. At the time, it was not clear just how much of the nuclear fuel had actually melted. We knew that much, if not most, of the fission products had been released from the fuel, but we also thought that a complete meltdown had not occurred. The various blue-ribbon panel reports agreed that *even if a complete meltdown had occurred, the amount of radioactivity released to the environment would not have been much larger* than was actually released. Later examination of the core showed that a large fraction of it, weighing several tons, had melted into a mass and dropped onto the bottom of the pressure vessel. For all practical purposes, a meltdown had occurred. TMI, therefore, was close to the worst possible casualty for

reactors of the type being built in America, yet its effects were a long way from the pessimistic speculations made before this information was available.

That is the most important information gained from the accident that the horrible scenarios envisioned in the worst-case "what if" studies are virtually impossible to create, even if one wanted to. This fact has not been fully assimilated into NRC's safety planning because it seems more conservative to go along with various grim but unrealistic hypotheses.

5. The total radioactivity released to the environment over the course of the accident was 2 to 13 million curies of the noble gases xenon and krypton, which are biologically inert, and 13 to 17 curies of iodine, which is damaging to the thyroid but short-lived (8-day half-life). No detectable quantities of other more hazardous isotopes such as longer-lived cesium and strontium were released. For every atom of iodine that escaped into the containment air, *five hundred thousand atoms stayed absorbed* in the water inside the containment structure. This was not the result of any particular action or procedure; it resulted from the fundamental properties of the materials and is a valuable safety feature of water-cooled reactors.

6. All of the investigators emphasized that they did not examine the broader question of the future of nuclear power, but noted that their findings did not "require the conclusion that nuclear power is inherently too dangerous to permit it to continue and expand as a form of power generation." Nonetheless, the accident was an extremely severe blow to the utility and to the worldwide nuclear power community as well. Because the public generally believed the hazard was much more severe than it was, and because of the shoddy performance of the officials and experts that the public was counting on explaining what was going on and protecting them from harm, the credibility of the industry and its regulators was seriously damaged. Utilities were no longer willing to jump onto the nuclear bandwagon, correctly perceiving that the public and

the regulators were going to be even more difficult to deal with from that time on. It is much easier for a utility to build no more new plants unless they have to, and those they build can burn oil or gas. By doing this, they minimize financially disastrous regulatory delays, and let someone else worry about the diminishing fuel reserves.

Findings and Recommendations

The various investigating groups were in general agreement as to deficiencies identified and recommendations proposed. There was plenty of blame to go around. Every organization directly involved with TMI performed at least some of their functions miserably. Yet, it is clear that this was in large part because they were let down by other organizations and individuals they were counting on. This should be kept in mind as I turn the spotlight on each party in turn. It should also be noted that a number of the investigators remarked that there was no reason to believe that the TMI plant was uniquely poor in any of these regards. In other words, an accident of this magnitude could have happened to a number of other plants.

1. The most obvious culprits were *the operators on duty* at the time. At several crucial points, they took actions that turned a trivial problem into a nightmare. If they had shut the block valve on the PORV, if they had not thwarted the high-pressure injection pumps trying to replace the lost water, if they had not shut off the main coolant pumps, if…if…if…. They clearly did not understand what was happening and were unprepared to deal with it. But none of the investigative groups concluded that the operators themselves were the main problem, so we have to look further. (For example, the control room design was much too complicated and poorly organized to permit operators to perform emergency functions effectively. The utility company did not design the plant, but it specified what it wanted, approved the design, and assumed responsibility for running it. Like most operating organizations, the utility was

too ready to delegate design responsibility to its contractors and assume they would do the job well.)

2. *Plant management* was not sufficiently in control of the daily operating situation. This showed up in several crucial ways:

   - Operators were told not to deviate from approved procedures, but there were no procedures for the situation that suddenly confronted them.
   - There were deficiencies in the approval and revision of procedures so that procedures in use had errors and ambiguities that operators had to work around. There was no routine check of plant status and valve line-up at the beginning of each work shift.
   - The maintenance situation was similarly deficient. Malfunctioning equipment was not promptly fixed. A supervisor testified that there had never been less than 52 alarms lit, which operators were told to ignore, a situation not uncommon in the industry at that time. But that would never be tolerated in the naval program.
   - Training, from top management on down, was inadequate. No one on site was able to diagnose what was happening and what to do about it.
   - There were no effective measures to apply lessons from other plants. Management had been given written reports from designers, consultants and operators, warning of problems. Some of these lessons were directly applicable to the TMI accident; but many of them had not been acted upon.

3. The *Nuclear Regulatory Commission* was roundly criticized by all investigators for several serious shortcomings:

- It viewed safety regulation primarily as a legal function, rather than a technical function. This had two serious implications: First, the emphasis was on meeting increasingly detailed regulations, rather than on defining and measuring performance against safety goals. Second, the Commissioners exempted themselves from participating in safety reviews of actual plants. But they also failed to provide adequate policy direction and central management of the agency.

- It carried out its technical functions in a highly theoretical way, and few of its people had plant-operating experience.

- It focused almost exclusively on an abstract “maximum credible accident,” and was not prepared to deal with lesser problems.

4. *The media*—about 400 strong—flocked to TMI like vultures to road-kill. When evaluating the public relations aspects of the accident, the groups that investigated the accident afterward generally questioned only whether the needs of the media had been met. They wanted assurance that neither the government nor the utility had held back or censored any information. The press properly reported that the handling of the incident was chaotic. But in many cases, the media generated irresponsibly inflammatory coverage.

- “Catastrophe” and “disaster” were commonly used to describe the situation. “Deadly” and “radiation” were as inseparable as “damn” and “Yankee” in the South. Television news programs used the sounds of radiation counters clicking and funeral dirges as background music.

- A newsreel cameraman, unable to find scenes of desolation, chased away bystanders and put up FOR SALE signs he had brought along for the purpose.

- The good, gray *New York Times* ran a story that read like a voodoo curse, conjuring up fearful Jungian archetypes: "A frightening array of biological problems in animals ranging from cats to cows ... spontaneous abortions, stillbirths, sterility, mutant offspring, blindness, defective bone structure and sudden death ... wild birds, game animals and snakes have greatly diminished in numbers ... One Hershey woman chose to have an abortion and then had herself sterilized rather than rear an infant where 'it will never be clean.'" (March 27, 1980). Such prose cannot be considered either a news story by an investigative reporter or a technical evaluation by an objective scientist. It was simply a political pitch written by a full-time professional anti-nuclear lobbyist.
- Anne D. Trunk, a resident of Middletown, Pennsylvania, and one of the appointed authors of the Report to the President, put the following statement into that report, affirming that it "represents the feeling of the undersigned and a majority of her circle of citizens who lived through the TMI accident." She disagreed with the report's finding that "the press did a creditable ("more reassuring than alarming") job of news coverage." She went on to say:

  > *In fact, these conclusions are not generally supported by the staff reports...too much emphasis was placed on the "what if" rather than the "what is." As a result, the public was pulled into a state of terror, of psychological stress.... the major networks proved to be the most depressing, the most terrifying. Confusion cannot explain away the mismanagement of a news event of this magnitude.*

5. The NRC expected that the utility would run the plant properly. The utility expected that their contractors would design and build it right. The operators expected that their training and procedures would tell them what they would need to know in an emergency. The media expected they would get a straight

story from the responsible officials, and these officials expected that the press would report the facts, not baseless speculations. These expectations were based on the fact that several hundred nuclear power plants had run for many years without any serious problems. Yet all these people were let down.

This brings us to a few very important questions:

What has changed since TMI? Are we any better off today?

Does all this just prove that people and machines are too unreliable to entrust with something as dangerous as nuclear power?

Changes Since 1979

Billions of dollars were spent to bring every nuclear power plant into compliance with every major and minor revision suggested in the TMI review. Cost was not a consideration. Two types of changes were needed: technical and organizational. Technical changes were aimed at ensuring that operators would have the information they needed and the training to do what was necessary. The aim was to improve the operators' ability to deal with any unusual situation that might arise, and thus to prevent accidents from occurring, or failing that, to minimize the impact of whatever might happen. The ability to accomplish such changes effectively depended on the organizational changes that were also implemented.

The second technical task was to take the data from the accident on fuel behavior, release of fission products, radiation levels at various locations, etc., to develop more realistic estimates of the impact of any future accident. Prior to 1979, no actual "field data" on a nuclear accident existed; that is, there were laboratory tests, some of them fairly large scale, and there were lots of calculations and computer simulations, but there had been no actual plant casualty. In that situation, it was felt that the conservative course was to assume each step in the process might be a lot worse than calculated, which in turn aggravated all subsequent steps.

However, the safest course of action does not usually result from compounding unduly pessimistic assumptions. For example, if you assume pessimistically that the air will be full of long-lived fission products and this condition will last for a long time, you would tell people to leave the area. If in fact, the radioactivity lasts only a few hours, it would be safer to have people stay indoors, rather than mill about on the highways. And if you know that the radioactivity released will not be enough to cause any significant public or environmental hazard, you should be prepared to determine whether this is the case and inform people as quickly as possible to alleviate any groundless fear and panicky actions that might otherwise result.

It was not difficult to state what kind of changes needed to be made. But to bring it all off required a basic change of attitude or mindset, and also a great deal of work, on the part of many thousands of people. The Naval Reactors program had shown that American industry could produce quality equipment, and sailors could be trained to operate it safely and reliably. But the TMI accident showed that you can't count on this happening time after time, without heavy management involvement and control, and the rigorous selection and training of personnel that made it work for Rickover. The NRC was part of the problem, but the lion's share of the responsibility for correcting the situation lay with the industry, which included more than fifty independent electric utility companies that generate and transmit electricity, and the scores of organizations and their extensive network of suppliers and subcontractors involved in designing, manufacturing, and constructing nuclear plant equipment

The Industry Response

The industry was quick to realize the importance of responding effectively to the TMI accident. Not only the blow to public confidence, but also the humbling experience itself made it clear that drastic action was necessary. A year after the accident, the industry was able to report that the following actions had been taken:

- A new research center, called the Nuclear Safety Analysis Center (NSAC), was established and staffed to study reactor safety questions. This was to work with the already existing Electric Power Research Institute (EPRI) that continues to provide generic research data to the industry. NSAC's functions were later taken over by EPRI, and NSAC was discontinued.

- A new organization called the Institute of Nuclear Power Operations (INPO) was set up to establish "benchmarks of excellence" and monitor the state of training, maintenance and operation of all of the nation's commercial nuclear power plants. INPO was not created to supplant the regulatory role of the NRC, but to provide a means whereby the industry itself, by acting collectively, could make its nuclear operations safer and more reliable.
- The information and analyses of the TMI accident were coordinated and published in a series of reports.
- Industry comments on various NRC proposals were coordinated to facilitate needed changes in response to the accident.
- A model emergency response plan was developed that could be used as a basis for individual plant programs. This plan provides for a "ready reserve" of personnel and equipment immediately available when needed and a generic public information plan designed to avoid the confusion seen at TMI.
- An industry-wide system was set up within INPO to analyze each and every abnormal occurrence. (NRC already required such events to be reported.) Any lessons or corrective actions gleaned from these events are promptly written up and distributed to all of the utilities to whom the lessons apply.
- A new insurance company, Nuclear Electric Insurance Limited, was created to provide partial protection from the financial impact of an accident, particularly the cost of buying

replacement power, which was the major financial cost of the TMI accident.

Views of a Political Science Professor

Joseph V. Rees, political science professor at the University of North Carolina at Chapel Hill, observing the tenth anniversary of the TMI accident, began to wonder and examined what changes had taken place since that accident to improve the regulation of nuclear plant safety. He is an expert on regulation of the workplace and on occupational safety. In his research, he was surprised to find indications that a great deal of reliance was being placed on the Institute of Nuclear Operations (INPO), an industry group. Noting that self-regulation of an industry is generally viewed with great skepticism, he dug in further. He found nearly unanimous praise within the industry for the effectiveness of INPO's operations, but this did not surprise him. Nor was it enough to convince him.

He then "systematically sought out INPO's strongest critics" in various anti-nuclear groups such as the Union of Concerned Scientists, Nader's Critical Mass Energy Project, the Natural Resources Defense Council, and the Nuclear Information and Resource Service. There he did find surprises. "They did not dismiss INPO's regulatory role as mere industry window-dressing. Far from it; indeed, one of the most knowledgeable and frequently quoted critics of this country's nuclear regulatory system, Robert Pollard, a former NRC inspector now with the Union of Concerned Scientists, thought highly enough of INPO to suggest (only half-joking) that the federal government should nationalize INPO and disband the NRC.... This was astonishing to hear, not only because it came from a leading industry critic, but because it stands in stark contrast to the prevailing social science view, which has it that industry organizations like INPO are typically weak and ineffective regulators."

Rees devotes an entire book to explaining how INPO works, and why it works so well. The key, he decided, is in the phrase that gave his book its title: *Hostages of Each Other.* He quotes Walter J. McCarthy, then-Chairman of INPO's board of directors and CEO of the Detroit

Edison utility company, addressing his fellow nuclear electric utility CEOs:

> It took the shock of a world-focusing event such as TMI to make us realize major changes must be made—in operations, in information exchange, in training, in management, in attitude, in culture overall. For the first time, I believe it hit us that an event at a nuclear plant anywhere in our country could and would affect each nuclear plant....Each licensee is a hostage of every other licensee.

Changes at the Nuclear Regulatory Commission (NRC)

The major recommendations of the various investigating groups regarding the NRC concerned the lack of management and policy definition by the five commissioners with equal authority who run it; its obsession with an abstraction called the "maximum credible accident," and its emphasis on the legal niceties of licensing procedures rather than on safety. Some of the reports noted that many of the NRC's practices actually worked against safety. For example, the long delays that resulted whenever a utility raised a new safety question provided a strong incentive for utilities to keep such concerns to themselves and concentrate on satisfying only the questions raised by the NRC.

The remedial actions taken by the NRC after the accident corrected some but not all of these problems.

- The headquarters operation had been in eight different buildings, with the commissioners not even in the same state with most of the staff (see Fig. 10.5). Now all the Commissioners and most of the staff are in two new adjacent buildings in Maryland.
- The commissioners voted four to one against going to a single administrator or even giving their chairman greater power. Their concern seemed to be that if this single administrator was perceived to be "pro-nuclear," the agency would lose credibility. (Such a concern seems to be unique to nuclear

agencies; you don't hear the head of the FAA worrying about being perceived as "pro-aviation.")

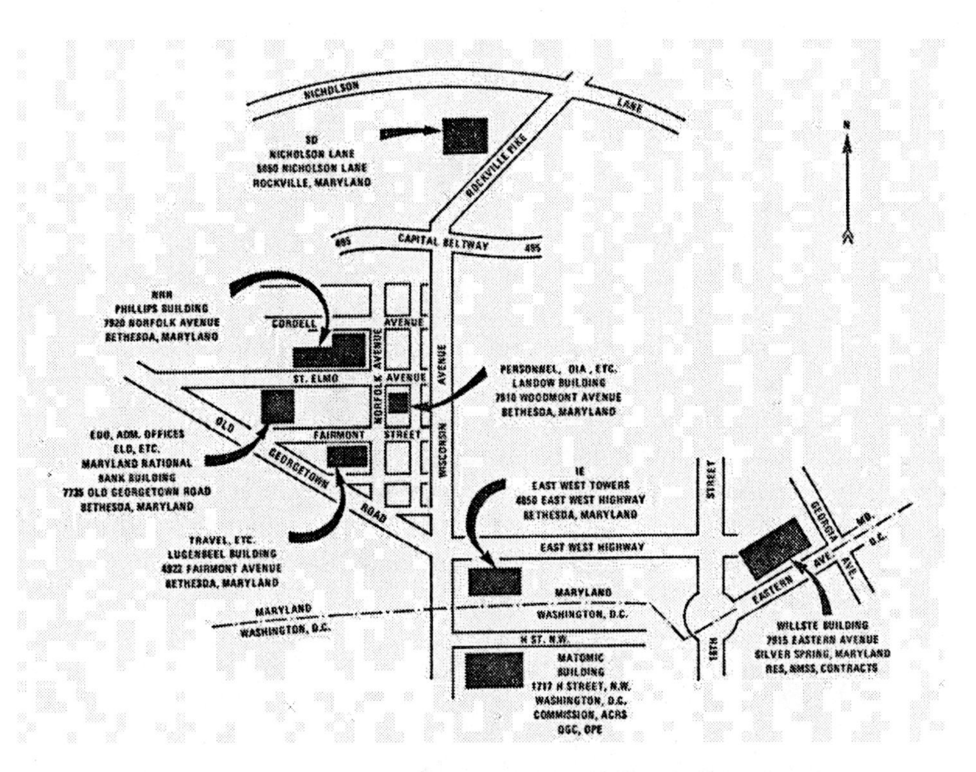

Figure 10.5 Locations of the various NRC offices (*Author*)

- Progress in orienting the agency toward a less legalistic view of their function was slower to come. There are a large number of technical people in the agency, but their experience is with analysis—"number crunching"—not operation. The focus on detailed mathematical analysis of improbable hypothetical scenarios with extreme consequences still prevails and supports work on evacuation procedures and other measures of questionable value in the real world. Even the work on realistic bits of the problem—probability of particular sequences of incorrect valve settings or component failures—are handled, in most cases, as challenges in statistical analysis. However, this situation has been improving during the last few years.

- The agency has passed some of its responsibilities for non-safety matters such as control of export licenses and antitrust determinations to other agencies.
- In its dazzling new twin skyscrapers in Rockville, Maryland, the NRC has installed an impressive system of equipment and procedures for receiving, displaying, processing and disseminating information, looking for all the world like NASA's operations center at Houston. If this works as advertised, and as I have seen it demonstrated, it should eliminate most of the communications foul-ups of TMI. Verbal reports of plant status are received and displayed in real time, (that is, as fast as they are transmitted), along with any requested plant meter readings of temperature, coolant flow, etc. As at Houston, technical people are assembled around the displays and they can calculate, make predictions and request further information. There are booths for representatives of the various federal, state and local officials, news media and various NRC representatives. There are also lines of communication into the system and out through the Internet to the public.

Do These Changes Solve the Problem?

Today most of the necessary structural elements are in place to support the high standard of technical excellence that results in safe and reliable performance. And the plants are showing the effects of this situation: unplanned outages are down, quantities of radioactive wastes produced each year are decreasing, and the number of days of uninterrupted power operation continue to rise. In 1991, TMI-1 was listed as the best of some 400-plus commercial nuclear power stations worldwide in terms of reliable on-line operation. Both in terms of plant design and operator competence, a significant improvement has been achieved. In particular, the existence of INPO as an industry-wide watchdog is proving to be effective. As previously discussed, the next generation of plant designs now available, Advanced Light Water Reactors (ALWR), promises even greater safety.

But no commercial plant that I know of ever reached the degree of knowledgeable management involvement that characterized the Naval Reactors Program, and I think it is inevitable that as the humbling experience of TMI fades into history, this involvement will continue to decrease. We must assume that this will bring with it some gradual lowering of standards until the next humbling experience. But we've also seen that the catastrophic vision of the impact of such an incident is wholly fictional.

Is Nuclear Power Too Dangerous for Humans?

The two most important points so far:

- As a result of TMI, the industry and the government have put in place needed organizations and procedures to set standards, to monitor and control operations and to exchange information. In addition, technical improvements have been made in plant designs, training and operating procedures. These actions should all significantly improve safety by reducing the chances of another accident and by decreasing the consequences of any accident that might occur. Evidence that this is working is shown by the operating statistics: all of the criteria by which one judges the quality of operating are continuously improving (see Figures 10.6 and 10.7).
- Nevertheless, people being what they are, eventually a combination of goofs may well lead to another accident of some kind. At least, that is the conservative assumption that we make.

In view of this, is it reasonable to proceed with nuclear power? This question comes up only because nuclear power is perceived by many people to present a potential for a catastrophe of unprecedented magnitude.

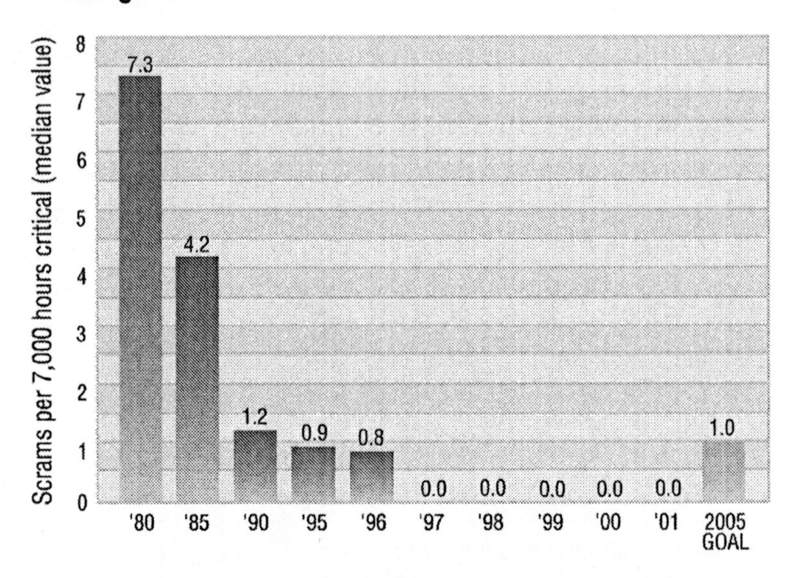

Figure 10.6 Number of unanticipated automatic shutdowns "per year" (*Nuclear Energy Institute*)

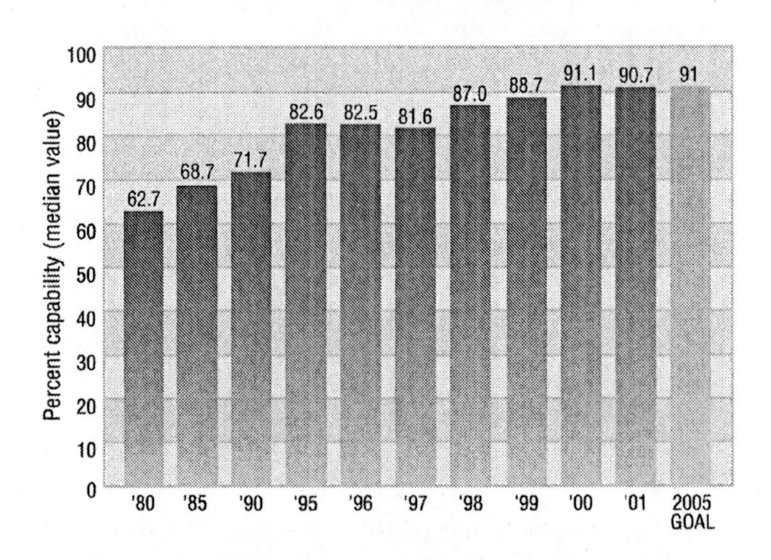

Figure 10.7 Percentage of plant capacity achieved (*Nuclear Energy Institute*)

It is seen to be almost like the dreaded California earthquake, the big one that will come some day and wreak havoc from Los Angeles to Chicago. *We can't stop the earthquake, but, by golly, we can sure stop building nuclear power plants.*

This perception was created largely by the atomic industry itself, in its efforts to examine every possible accident scenario, no matter how fanciful, and take the worst value for each of a series of hypothesized events.

The Bottom Line

The hypothesized nuclear accidents were indeed scary, but now we have hundreds of real reactors that have been operating safely and reliably since before the birth of most of the world's living population. And we have thoroughly studied the results of the only applicable real-world nuclear plant accident, the one at TMI-2. These studies tell us that if you somehow let a third to half of the reactor core melt down and fall onto the bottom of the reactor vessel, the consequences outside the plant boundary are trivial.

It is possible to imagine further problems: assume *all* the core melts; assume the containment vessel integrity is somehow breached. Even in the worst imaginable case, the maximum possible number of casualties we can foresee is less than the number of deaths and injuries that can result from a single day's automobile accidents in the United States or the release of chlorine gas from a single breached railway tank car.

The U.S. Nuclear Regulatory Commission has conservatively recognized this situation by revising the "source term" that utilities must use in applying for a license for a nuclear plant. The source term is the amount of radioactivity assumed to be in the containment atmosphere as a result of a complete meltdown of a reactor. The NRC assumes that this radioactivity somehow escapes from the reactor vessel and leaks from the containment vessel into the environment at the maximum leak rate for which the containment was designed and periodically tested. Even with these pessimistic assumptions, the calculated effects on the public if no remedial action is taken is shown to be comparable to many other minor casualties we face each year.

The Data from Karlsruhe

The German Federal Ministry of Research and Technology has been carrying out a massive theoretical and experimental study of reactor safety at a laboratory in Karlsruhe for nearly twenty years. Phase A of the study was completed just about the time of the TMI accident. This study consisted of independent theoretical analyses of all conceivable events that could lead to a nuclear accident, with estimates of the probability of each and estimates of the expected consequences of such an accident. The numbers arrived at agreed quite well with those of the American safety studies.

Phase B of the program was reported out in 1990. In addition to further refinements in the theoretical calculations, the study reported some large-scale experimental work. At the Sascha melting facility, large quantities of "corium" (uranium oxide, with other chemicals added to correspond to the fission products) were melted to measure release rates of the various fission products out of the molten fuel. At the Beta facility, over half a ton of molten fuel was poured into concrete crucibles and kept molten by electric inductive heaters, to test how concrete structure below the pressure vessel might react with the fuel. (The evidence from TMI showed that fuel escaping from the vessel was highly unlikely). The Demona experiments were to test the behavior of fission products escaping from the fuel into the containment building; to determine which elements plated out on the structural surfaces, which ones dissolved in the water, which reacted chemically with other elements, and which stayed in the air, free to leak out of the containment if a leak developed.

These tests were about as realistic a large-scale mock-up of various aspects of a nuclear meltdown as one could build. The program demonstrated that "the consequences of severe accidents at German LWRs (Light Water Reactors) have been overestimated in earlier calculations by perhaps one or two orders of magnitude, i.e., ten-fold or a hundred-fold:"

> ... the consequences of even the worst core melt sequence need no longer be considered as a national catastrophe.... Therefore no necessity exists to consider any so-called advanced safety concepts which are predominantly based on the (untested) principles of passive features and inherent safety characteristics ... the researchers believe they have assessed all the major possible accident sequences in German LWRs.

The Most Important Fact about Nuclear Power Safety.

When we put together all the hard data we have gathered from the Three Mile Island accident and from the massive theoretical and experimental programs since that time, the most important fact we have learned is that the notion of a catastrophic impact on the public or on the environment is shown to be technically unachievable, even if no effective remedial actions are taken. Nuclear power plants of the kind now operating in America, Western Europe, and the Pacific Rim are simply not the unprecedented hazard many have feared.

That is the bottom line. I have described aspects of nuclear plant design and operation that I am not happy with: I believe that the NRC's legalistic and theoretical approach to safety has been wrong-headed. I think many of the design features imposed during safety reviews have not added significantly to safety. (There are good indications that NRC's thinking may be improving significantly, but we won't know for sure until licensing review is initiated for a new reactor.) And I think that the next generation of plant designs, ready for construction, is considerably better from a safety standpoint than those now operating. Even so, *I am wholly convinced that the commercial nuclear plants now operating are as safe or safer than any other means of producing electric power.* I base this conclusion not on faith in the infallibility of human operators or mechanical safety devices, but on the inherent physical laws of nature that limit what can happen.

It is, of course, highly desirable to avoid another costly nuclear plant accident, and extraordinary steps have been taken to accomplish this. I believe that any reasonable person would agree that these precautions and procedures will indeed make the reoccurrence of any kind of serious nuclear accident nearly impossible. But even if we fail in that endeavor, we can expect that the resultant impact from an accident on public health or on the environment will be negligible compared to other risks we face routinely every day. In particular, the risks of a nuclear power plant accident are small compared with the inherently lethal effects of coal-burning plants, which we continue to accept as the price for getting reliable electricity.

The Importance of Energy

We often hear that *conservation* is the answer to the world's energy problems: *Just don't use so much energy, and the problems won't arise.* But this is really more of a dodge than a solution. Most of the material things that need doing in the world require energy. And our need for energy is increasing exponentially, as the rest of the world wants some of the things we have. Conservation will lower the rate of increase somewhat, and to the extent we can do that, we should. No reasonable person is in favor of waste. But conservation merely puts off the day when we reach any given crisis point. Before long, each of these crisis points will be reached, with or without conservation, and then we'd better have a solution ready.

Our fossil fuels—coal, oil, and gas—are running out. Sixty years ago I was told we had 25 years of oil left, but more intensive exploration has kept extending that deadline. However, like conservation, intensive exploration and mining just put off the day of reckoning. Those fuels that supply most of our energy today are running out, and we will have to have an alternate source of energy.

A glance at the problems confronting us shows why our energy needs will continue to grow. We need to build decent housing, highways, railroads and airports—all over the world. We need to clean up contaminated waterways and build wastewater treatment plants and drinking water purification systems. There are contaminated sites that need detoxification. There are places where the air needs cleaning. Hospitals, schools, libraries, sports arenas, art galleries, theaters and museums should be built. City streets and buildings need light, heat and air conditioning, and rural areas need electrification. Wells must be dug, and where there is not sufficient ground water, we must desalinate seawater (which is in essentially unlimited supply). Then we need to pump the water to wherever it is needed. Whether Americans or others supply these needs, and whether the load falls on governments or on private individuals, whoever does it will need energy and lots of it.

The burst of conservation that followed the oil crunch of 1973 was effective; total energy consumption in the U.S. dropped markedly. But *electricity* usage surged upward because much of the energy saving was achieved by switching to electricity from older, less efficient processes—for example, in steel-making. Energy is a tool to achieve various goals. It makes no more sense to set energy reduction as a priority than it would to try to restrict the use of hammers or computers. Energy use, like language or fire, is neither good nor evil. It is perhaps the most basic tool we have, and it is up to us to use it wisely and effectively.

Although wood and fossil fuels are fast being used up, we are blessed in having discovered just in time how to harness for our own use the virtually limitless energy of the atomic nucleus.

Figure 11.1 The Environmentalist looking for river polluters (above) and in his day job, working on TMI cleanup (*Janet Cole*)

11. The Environmentalist

In my personal odyssey, a recurring theme is the very special people I have met along the way. There is Dr. Alvin Radkowsky, who was Admiral Rickover's Chief Physicist in developing the first nuclear power plants. He was also an ordained orthodox rabbi with a Vilna Talmud, who spent many a Friday night and all day Saturday sitting in airports when his plane was delayed, because his religion forbids travel of any sort on the Sabbath (which lasts from sunset Friday to sunset Saturday). There is Bill Spargo, a white-hat sailor who got master's and doctor's degrees in medical engineering in about four and a half years total and became a nuclear plant designer. You could also find him, resplendent in tuxedo, chairing the monthly meetings of the 150-year old Washington Philosophical Society. And there's the registered professional electrical engineer who is also a card-carrying member of the International Brotherhood of Electrical Workers; the future submarine captain who had to interrupt his career for an extended holy pilgrimage to Mecca, and many others. I could fill more than a book, just talking about these fascinating people. But here I will tell you about just one, a close friend, recently deceased.

Noman Cole was used to being called "Norman" by new acquaintances. The pronunciation, with or without the "r," is not too different along Pohick Creek and the Potomac River in Virginia where he lived, and he admitted the old family name was unusual. In the early spring of 1970, he was again out at dawn in his little boat, moving slowly up the dirty Potomac River, watching for the telltale signs of polluters surreptitiously dumping wastes into the deteriorating river. He had no authority for this one-man crusade, only his vice-presidency of the Mason

Neck Citizens Association. But he had the help of his own version of Sherlock Holmes' Baker Street Irregulars, river people who came to him with tales of midnight dumping and other illicit activities. On any pretext he would talk to you about what he was trying to do.

Noman and his wife Janet were an unusually attractive couple, in both appearance and personality. His slightly wavy, brush-cut hair, ready smile, and strong, confident voice, conveyed a youthful vigor. Janet has the cool beauty and chiseled profile of a latter-day Katharine Hepburn, and the confidence that comes with having proved that she could make considerable money by shrewd investment and management of their earnings. You couldn't help but enjoy talking with them in their home on the Potomac River, or weekends at their 300-acre stock-farm with their neighbors making apple butter the old-fashioned way in a blackened copper pot over an open hickory fire. You were apt to find yourself nodding in agreement with whatever point they were making. Articulate and persuasive, they were effective activists for a number of community and environmental causes. The soft Florida accent Noman grew up with sounded perfectly at home in the rolling Virginia countryside.

Noman was born March 10, 1933, in the South Carolina "low country," son of a Navy flier. He told me of his memories of going quail hunting and fishing with his father in his early years. "*Very* early years," he emphasized, "because the family moved to northern Florida when I was just four, and I still remember Carolina." The Coles picked a place on a tributary of the Saint Johns River near Jacksonville, and rivers and lakes continued to dominate their life.

"Ever since I can remember, rivers, lakes, bays and inlets have been almost sacred to me," he said. "I just couldn't imagine how anyone could willfully pollute a waterway. I remember one time as a kid, being out in a little boat, marveling at the fish and plants in the water, when suddenly I came to a big, black cloud in the water. All around, there were no more fish or plants. Somebody had dumped sludge of some kind, and I felt as if my own personal life had been fouled. You know what I mean? This was before it was chic to be 'environmental.' But my feelings on these things never died down or went away as I got older.

"There is no excuse for people dumping huge quantities of noxious wastes into a river," he argued forcefully, poking his finger into my chest. "And there is no mystery about how to stop it. We have plenty of laws, and they clearly prohibit what is going on. All we have to do is act. Once it becomes clear that people will not tolerate pollution of their public waters, it will stop. It's not a complicated problem."

"If it's so easy to stop, then why does it continue?" I asked him. He grabbed my arm and pulled me in close, making it clear that he was about to say something important.

"It requires politicians to act," he said simply. "Politicians are quite willing to act where they know the people demand it. So far, we haven't built up a grassroots demand for cleaning up the river. People are demanding lots of other things of their politicians, and those other things are what get done. But that can be changed. After all, nobody is in favor of pollution."

"How about the big companies that do the polluting?"

"You'd be surprised. Big corporations are a minor part of the problem. The biggest offenders are governments: federal, states, counties, cities. Sewage treatment plants. And hospitals. It's funny; people think of corporations as the big enemy on issues like this, and people look to governments to protect them. But the fact is that government employees are just as human, and are just as intent on protecting their own institutional turf as the people in corporations. But I'm going after all of 'em—whoever they are. The tons of crud being dumped into the Potomac have been increasing by leaps and bounds every year. There's no reason that can't be reversed. I want to see the Potomac back to assimilative levels in time for the Bicentennial Celebration in 1976. *Assimilative levels*—that's a term we use to describe a condition where the natural cleansing processes in the river can keep up with the waste input. I want to see people catching edible fish again. Do you know that fish have returned to the Thames River, in England, for the first time in two hundred years? We can do that too."

Noman talked in the slow, soft, Florida way, giving the impression of being laid-back and southern, but his body betrayed a lot of stored-up energy. In the office he would swing around sideways in his secretarial-type desk chair, draping an arm over the back, occasionally hanging one leg over the chair arm. But his arms would begin to flail a bit, and his hands wave, point, open and then shut again. He stroked his chin, brought his forefinger across his mouth for a moment, then massaged his neck thoughtfully, and then, suddenly, one hand was gripping the arm of the chair, relaxing, gripping, then it was quietly in his lap. At times he folded his arms across his chest and looked thoughtfully out the window as he talked.

He applied that energy effectively in high school to win the half-mile race at the state track meet, and with a partial scholarship in track he enrolled in engineering at the University of Florida in Gainesville. During his last year in high school he read about Jacques Cousteau's "aqualung" invention, forerunner of the SCUBA diving gear. There were pictures and diagrams in such exciting journals as *Popular Science* and *Popular Mechanix*. He and his young friends could see that the critical element in the system was the regulating valve that brought the air pressure down from the 2,000-plus pounds per square inch in the tanks to a safe pressure just high enough to flow into the lungs but not so high as to endanger them.

All the local high school kids knew there was a dump for old airplanes outside of town, and they knew that pilots sometimes needed to sip air from high-pressure air tanks when they flew at high altitudes. The kids ran to the dump, and after a little scrounging—sure enough, there were some old regulating valves! In no time, the boys rigged up masks and built themselves amateur SCUBA gear. These tested out fine in the swimming pool, but when they took them to sea, they found the regulator valves began to fizz like Alka-Seltzer tablets. They didn't realize that the magnesium alloy valves were dissolving in the seawater. "But God protects drunks and little boys," said Noman, and somehow the high-pressure air systems did not fail disastrously. So, armed with spearguns they made from CO_2 fire extinguishers, they made the underwater world a playground whose beauty continually filled them with reverential awe.

In college, Noman spent a lot of time studying power plants. Those were the days when "smoke-stacks" pouring soot and carcinogens out onto the countryside were looked at fondly as evidence of prosperity, but Noman recalled being appalled at the quantity of black smoke that poured uncontrolled from the coal-burning power plants. He found that even the oil-burning plants were apparently the cause of deterioration of his mother's stockings. Like many of his fellow-students in that sunny state, he turned hopefully to a study of solar power. But he came away convinced (to his disappointment) that the heat of the sun—hot as it was on his back—was too dilute to be a promising replacement for coal or oil.

"It was easy to calculate that, even if you captured every bit of sunlight that fell, and converted it to useful energy in a highly efficient way, it would still take thousands of acres of solar collectors to match the output of one of those big power-generating stations," he recalled regretfully. "Then what do you do when it's dark and cloudy for days at a time? That happens, you know, even in Florida. You have to have enough generating capacity to run without the solar plant. And afterwards, when the sun shines again, you have that stand-by plant standing idle, but still paying taxes, insurance and maintenance. That doesn't make sense."

In the early 1970s, Cole and his Virginia neighbors became increasingly concerned with the algae that encroached farther every year into the river. Rumors started circulating about the huge new wastewater treatment plant that was to empty into Pohick Creek, which in turn empties into the Potomac just upstream of Mason Neck. It was said that the plant would have a large by-pass line that would dump raw sewage into the creek any time there was trouble in the plant. Cole sent for the complete piping drawings of the plants, scrutinized them carefully and found no by-pass. That should have been the end of it, but a neighbor persuaded him to go look at the plant itself, and there it was: a huge, four-foot-diameter by-pass line. "That's what lit the rocket," said Cole.

Cole started checking other details in the plant that seemed satisfactory on the drawings, and he found further problems. The drawings showed two independent sources of electric power coming into the plant, which was an essential safeguard. But it didn't show that the two power lines ran along the same poles for the last part of their journey. A truck or a car or lightning striking one pole could take out both lines at once, leaving the plant dead and dark. At the urging of neighbors, Cole then looked into some of the other nearby sewage plants and found they were also deficient in many ways, both with regard to piping and other mechanical features, and also in some of the basic chemistry being used. In the course of this work, he soon became comfortable working with zoning regulations and local politics.

With pressure from the local citizenry, some of the plants were upgraded. Everyone was amazed at the marked results achievable with modest plant changes. The power lines were strung on separate poles, and a truly separate emergency system was created at very little cost. No effort had yet been made to remove phosphorus, but when this was done, the algae problem diminished precipitously. By proper use of iron salts, the efficiency of the other removal processes was also greatly enhanced.

About that time Linwood Holton was running for Governor of Virginia on what we would now call a "green" platform, and in January 1970, he took office. He revitalized the moribund Water Control Board, and when a neighbor put up Cole's name for membership, Cole was appointed although he had never met Holton. There were four new members on the board and three holdovers. The Chairman resigned and Cole was made Chairman, right from the start. Stories of Cole's successes had gotten around, and the other new members were gung ho to do big things. The holdover members were naturally resistant at first, but Cole was impressed with their honesty and their good intentions, and before long he had won them over.

Cole found more and more public support for his aggressive defense of the river. The builders and developers, however, who fueled and profited by the rapid growth in the area, seemed to be a natural enemy. Since new waste treatment plants meant unpopular referenda for new taxes or new bond issues, developers have historically tried to hide the fact that new housing and new shopping malls lead to the need for new waste-treatment plants. Cole had to decide whether to take on this powerful lobby in a fight he might not be able to win.

Again, boldness paid off. Cole approached the developers as potential allies. He showed them you cannot fight the inevitable, that a given builder might get his project through, but this would only increase the pressure on the next project. He convinced the National Association of Home Builders that the public was increasingly concerned over pollution and ready to support funding for needed waste treatment facilities. "You guys can either be forced into going along, and look like bums, or you can lead the parade and look like heroes," he told them. They chose the hero's role and joined Cole in his crusade.

"Environmental Protection? Ha! I call 'em the Environmental *Pollution* Agency," said Cole. "All over the country, thousands of communities had begun to realize that they can't keep using their rivers as sewers, and they were ready to build new, modern waste treatment plants. Then the feds create the EPA who says, 'Wait a minute. Don't build your plant yet. Wait a few years until we figure out what our standards and regulations and procedures are going to be, and then if you meet them, we may give you some federal funds. And if you don't meet them, we may fine you and require you to modify your plant at great expense.' So all these communities, who were prepared to spend their own money to clean up their rivers, were stopped in their tracks, waiting for these clowns in Washington to figure out what their rules are going to be. And in the meantime, people continue to dump raw sewage into the rivers. This has been going on for years. Is that any way to protect the environment?!"

Cole not only talked, he wrote: letters, point papers, petitions, positions papers. And he wrote as he talked, with lots of underscoring, like *The Kiplinger Letter*. With audacity, eloquence and logic on his side, and the expected opposition firmly supporting him, Cole was able to convince Governor Holton to act vigorously. In 1972, *Washingtonian* magazine, in a well-publicized cover story, named Cole one of ten "Washingtonians of the Year, men and women who in our judgment have done the most to improve the quality of life in Washington." Alongside a picture of him at the helm of his trusty boat, they wrote: "If the Potomac River soon flows more cleanly, much of the credit will be due to the courage of an aggressive and effective environmentalist named Noman Cole ... he is directly responsible for action that by mid-1972 should reduce by half the tons of untreated sewage now poured into the Potomac daily....Coming from the grassroots, Cole is an outstanding example of what one man can do when his concern is backed by knowledge, courage, and determination."

In the years preceding Cole's term on the Board, pollution had been building up in the Potomac. The river began to stink; it was condemned for swimming and then for fishing. People complained about it, but they were resigned to the situation as an inevitable price of "progress." In the course of ferreting out places where pollution entered the river, Cole discovered "The Georgetown Gap," a 2,800-foot missing section of pipe in the two huge pipelines that carried raw sewage through Washington from the Maryland suburbs to the Blue Plains Sewage Treatment Plant. The good burghers of Georgetown, proud of its historic ambiance as part of pre-Revolutionary Washington, did not want their neighborhood torn up to bury a large pipe, so they diverted the raw sewage flow into the Potomac River near Fletcher's Boathouse, just upstream of Washington's new, multimillion dollar Kennedy Center for the Performing Arts.

The Georgetown Gap became a cause celèbre. Cole dug out the long hidden facts and brought them into focus and then into public scrutiny. He got Governor Holton to threaten to sue the State of Maryland, and even to sue the federal Environmental Protection Agency to enforce its own regulations. It was finally agreed that something had to be done, but nobody wanted to dig a huge trench across Georgetown. Finally, someone noticed the Whitehurst Freeway, an elevated express highway that runs parallel to Georgetown along the river, and it was realized that a jumper pipe could be slung beneath the highway without any significant disruption of the city. The job was done, and clarity, fish and finally swimming, returned to the river.

By 1974, Cole had completed his four-year term on the Water Control Board. Although the population using the river had grown explosively during that period, his efforts cut the input of solid pollutants 25-fold, a reduction of 96 percent! He continued on as a member of the state's Energy Advisory Council in his efforts to protect the environment.

The Fairfax County Chamber of Commerce gave him their "Man of the Year" award. The internationally read *Washington Post* cited his work in an editorial, as an inspiration to those who become tired and cynical as they survey the magnitude of the problems facing humanity. "Courage and competence, but above all ceaseless effort, does pay off," they wrote.

People who observed all this sometimes asked him how he found time to carry out all these efforts.

> "This isn't even my full-time job," he would reply. "I do this as a hobby."
>
> "Then what is your day job?" they'd ask.
>
> "I'm a nuclear power engineer—I work on nuclear power plants."

Noman couldn't suppress his annoyance that people found this contradictory. "They read about oil tankers polluting our coastlines and leakage from storage tanks fouling the earth. They know what strip-mining coal does to the earth, and they've been told about the 30,000 deaths each year from respiratory illnesses resulting from coal-burning power plants. And yet they seem to think that if you are pro-environment you have to be anti-nuclear. It just doesn't make any sense!"

Noman was an important player in the early Naval Reactors program and became a leading engineer in our engineering firm, MPR Associates, Inc. He lived intensively, enjoyed his work immensely, and died in the same mode, by skiing at top speed inadvertently into a tree.

Figure 12.1 Nuclear medicine is big news
(*Science News*)

12. The Other Ninety Percent

The Fiftieth Anniversary Party

November 16, 1992, was a big day for the nuclear power community. Everyone who could make it was at the McCormick Center Hotel in Chicago to celebrate the fiftieth anniversary of nuclear power, which started with the first controlled nuclear chain reaction carried out by Enrico Fermi and others in a squash court under Stagg Field Stadium at the University of Chicago. It was an indication of the youth of the nuclear pioneers who participated in that first fateful experiment that many of them were there fifty years later to celebrate the occasion.

It was also the sixtieth anniversary of the discovery of the neutron, the hundredth anniversary of the first commercial production of electricity at Edison's Pearl Street station in New York City, and the twenty-fifth anniversary of the first commercial production of nuclear power at the Shippingport station in Pennsylvania. I was there as the invited kick-off speaker at the Chairman's Special Session on Early History, and I had taken it upon myself to arrange a reunion of Rickover alumni for later in the day, so I was fully embroiled in the activities.

On the afternoon before the Dinner of Reminiscences, one of the important figures was not in the ersatz grandeur of the ballroom with the others. Seventy-one year old Rosalyn Yalow, Nobel laureate medical physicist, one of the founders of the new discipline of nuclear medicine, was outside in the cold. Although most of us in the ballroom didn't realize it, protesters from Greenpeace were picketing the hotel with placards demanding that the nuclear industry be shut down, and Dr. Yalow went out to talk with them. She is blunt and articulate, but she didn't go to harangue them. She found out they were serious in wanting a complete shutdown of the nuclear industry, so she said, "You must understand, that would close down the entire field of nuclear medicine." This announcement did not bring forth any indication of concern from the protesters, so Dr. Yalow gave them a challenge: "Are all of you willing to sign statements that you will forego forever, for yourselves and your loved ones, all use of diagnostic and therapeutic techniques involving radiation or radioactive materials? Because these materials would not be available, you know, once you had shut down the reactors." There was no immediate response to her challenge, and she merely said, "You'd better think about it. It's a serious matter. I assume you want to be responsible in your actions." And she left them to their thoughts.

That night, at the banquet, she challenged the rest of us. "This is not the American Nuclear Power Society," she said. "Nor is it the American Nuclear Weapons Society. It is more than that. It is simply the American Nuclear Society. And ninety percent of the people dealing with radiation and radioactivity are in nuclear medicine, industrial radiography, sterilization and other activities completely removed from the power business and the weapons program. It's time the members and officers of this society fully recognize that fact and take it into account."

The extent of this benign neglect was dramatized two years later, in 1995, when the New York Cornell Medical Center announced that it was being forced to curtail indefinitely all of its research that required use of long-lived radioisotopes. In fact, twelve of the twenty hospitals in New York state licensed to handle radioisotopes found themselves in the same position later that year. The reason is simply that the state and federal governments had not gotten around to agreeing on a storage site for the low-level radioactive wastes that these hospitals produce. More than ninety percent of these materials are just barely radioactive and will have decayed completely within thirty years. The eighteen different participating federal agencies were still quibbling among themselves sixteen years after legislation was passed requiring the government to provide facilities and standards for storing these low-level wastes.

Dr. Yalow was right. Like almost everyone else in the room, I had read about nuclear medicine and was vaguely aware of its importance, but its full impact on current medical practice had not sunk in. I also knew that radiation and radioactive materials had widespread use in industry. But I did not realize that of the 1.8 million jobs associated with nuclear energy, 1.6 million are outside the electricity-generating field, and bring in $110 billion in revenue each year. Fewer than five percent of the licenses issued by the Nuclear Regulatory Commission involve electric power production. Much of the low-level radioactive waste is produced by hospitals, not by nuclear power plants.

Patients are assured that the low-level radioactivity used in medical diagnostics will not harm them in any way. Yet when they walk out of the hospital, regulations require that they be treated as radioactive hazards. If they work in a controlled radiation area of a nuclear facility, they would be forbidden to return there because their internal radiation would show on their personal monitor as an impermissibly high dose. They cannot use bathrooms in the nuclear facility because these are monitored for radioactivity, and more than once such usage has led to a plant-wide search for the source.

This situation continues for several days, up to a month or so, depending on the half-life of the radioisotope used. For several days, the patients are not to sleep with their spouses or hold their children in their laps. Their urine and feces are considered to be "hazmat" (hazardous material), strictly regulated and controlled. Gowns and other materials used at the hospital are treated similarly. Storage and handling of all such "low-level waste" is burdened with such controls, regulations and reporting requirements that a number of hospitals have simply stopped providing these life-saving services. Such is the extent of our radiophobia.

Although I have had little personal experience with these applications of the nuclear technology, I feel I should address briefly "the other 90 percent." These depend on the use of radiation or radioactive materials, which in turn depend on the operation of nuclear reactors to create the radioactivity (although some are now produced by "accelerators," the atom smashers of science fiction lore). A sprinkling of examples will illustrate the extent and variety of uses.

Nuclear Medicine

Of the 30 million people who are hospitalized each year in the U.S., one in three is diagnosed or treated with nuclear medicine. More *than 10 million nuclear medicine procedures* and more than *100 million nuclear medicine tests* are performed each year in the U.S.

Many of these procedures, both diagnostic and therapeutic, depend on the tendency of certain elements to concentrate naturally in specific parts of the body: iodine in the thyroid, phosphorus in the bones, potassium in the muscles, boron in tumors. By injecting slightly radioactive isotopes of these elements into the body, doctors can trace various body functions in action. By using highly radioactive forms of these elements, the specific organ to which the element goes can be irradiated therapeutically.

Some of the procedures using this technology include:

- *Myocardial profusion imaging*, to map blood flow through the heart.
- *Bone scans*, to detect cancer 6-8 months earlier than X rays.
- *Lung scans*, to detect blood clots in the lungs.
- *Radioactive tracers "hooked" to antibodies*, a new diagnostic tool awaiting FDA approval.
- *Radioactive cobalt,* used to focus a sharp beam of gamma rays directly on a cancer to destroy it with little damage to adjacent tissue.
- The *"gamma knife,"* a highly focused radiation beam used in brain surgery to remove otherwise inoperable deep brain tumors.
- *Radioactive iodine,* proven so successful in treating thyroid diseases that it has virtually replaced thyroid surgery.

- Ninety-five percent of all *new drugs are tested with radioactive test procedures* prior to being approved for public use.
- Nearly every facet of modern medicine draws on nuclear medicine, from heart pacers powered by radioisotopes to *surgical tools, sutures, gloves and supplies routinely sterilized* by radiation.

Figure 12.2 lists some of the radioisotopes used for this work.

"Nuclear Medicine"
Molybdenum-99
Technetium-99
Iodine-131
Indium-111
Gallium-67
Iodine-123
Strontium-89
Chromium-51
Gadalinium-153
Cobalt 57
Xenon-133
Thallium-201
Phosphorus-32
Yttrium-90
Holmium-166
Fluorine-18
Carbon-11
Nitrogen-13
Oxygen-15
Rubidium-82

Sealed Sources
Cesium-137
Iridium-192
Strontium-90
Cobalt-60
Palladium-103
Iodine-125

Genetic Research
Phosphorus-32
Phosphorus-33
Sulphur-35
Hydrogen-3 (tritium)
Carbon-14

Figure 12.2 Some of the Radioisotopes Used for Medical Diagnosis and Treatment

Stimulating the Body's Defenses

The use of high doses of radiation directed at tumors to kill them is well known. Less widely appreciated are the uses of *low-dose irradiation* (LDI) to stimulate the immune system and other biological defense systems. In this procedure, the entire body, or sometimes just the thoracic region inside the rib cage, is irradiated. Typically, the patient would receive 10 to15 rad delivered in a minute or two, and this dose would be repeated two or three times a week for five weeks, giving a total for the series of 150 rad. This dose-rate is high enough to get the body's attention, but the total dose is low enough to cause no adverse effects.

Radiation doesn't have to produce more cellular damage than metabolism to get the body's attention. Its effects are differently distributed in the cell. Metabolism creates free radicals rather uniformly throughout the body. Radiation, on the other hand, distributes its energy in globs, which are more easily recognized by the body's defense mechanisms but no more damaging physiologically.

This process is illustrated in Figure 12.3. On the left is a computerized tomographic scan, a sort of x ray, of the skull of a patient from the top. The arrow points out a large tumor in the upper nasal cavity. A similar view on the right, taken eight weeks later, shows the tumor virtually gone after low-dose irradiation of the rib-cage. The tumor was untouched by the radiation itself; the tumor was destroyed by the body's natural defense system, enhanced by the stimulation of low-dose irradiation.

Fighting Infection

Only a few months after roentgen 's discovery of x rays, papers began appearing in scientific journals describing the use of low-dose irradiation to disinfect wounds and kill low-grade infection. The authors were aware that their primitive x-ray machines were not strong enough to kill the bacteria, and they surmised correctly that the radiation must be stimulating the body's immune system. Wounds where gas gangrene previously called for immediate amputation, resulting in only fifty percent survival, showed dramatic recovery without amputation after radiation treatment, and only a few percent mortality.

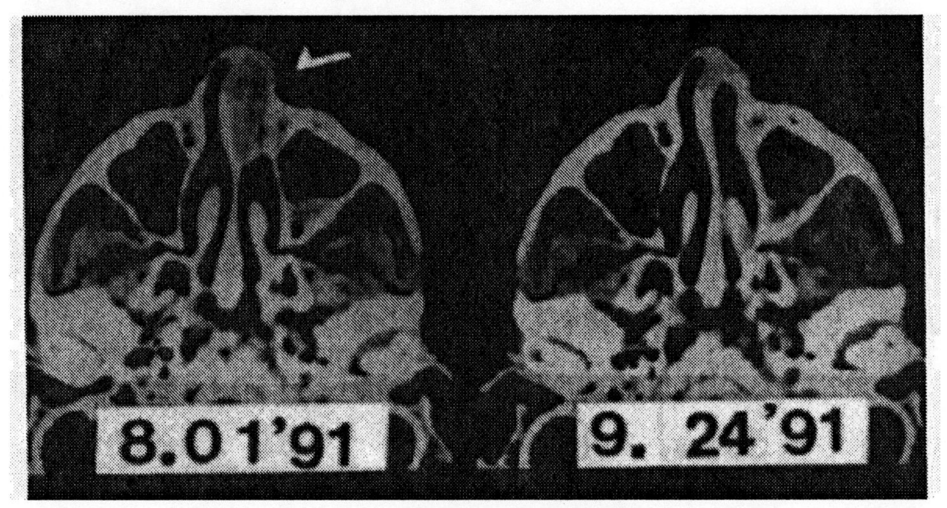

Figure 12.3 Tumor near brain cured by low-dose irradiation of rib cage (*Y. Tokai, Elsevier, 1992*)

This approach to a large number of syndromes was successful and widespread until the introduction of sulfa and other "miracle drugs" in the 1940s and '50s. After the War, aggressive promotion by pharmaceutical companies, assisted by a growing fear of radiation, led to drugs as virtually the only alternative to surgery, and the history of radiation treatment has been all but forgotten. Recently, however, the specter of an increasing number of drug-resistant pathogens makes irradiation worth a second look.

Industrial Uses of Radiation

Gamma radiation penetrates materials but loses some of its energy doing so. This leads to a large number of non-medical applications in industry. Sheets of paper, metal, plastics or other materials can be passed between a source of radiation and a radiation detector. The detector can rapidly and continuously **record and/or control the thickness,** just by measuring the amount of radiation that gets through. If, instead, the space between the radiation and the detector is filled with gas, the detector measures the density of the gas and thus becomes either a **pressure gauge** or a **vacuum gauge**. Variations on this process are used to **measure density of road surfaces** and subsurfaces**, packing of granular material, thickness of coatings, thickness of egg shells**, etc.

Gamma rays are, in effect, high-energy x rays and can be used to "x-ray" thick metal parts that would be completely opaque to ordinary x rays. This technique has become an indispensable tool for **inspecting metal parts** for weld defects, casting flaws, inclusions and other unseen imperfections that could pose a safety hazard if these flaws were not detected. The **rubber in radial tires is toughened** with radiation, and the alignment of their steel belts in checked radiographically. **Computer disks are irradiated** to make them "remember better," and **certain characteristics of textiles, such as ability to repel water**, are enhanced by radiation.

Geologists test the properties in boreholes by **measuring the natural radioactivity in the soil or rock**. This technique is particularly valuable in searching for oil. Devices lowered into the borehole continuously read out the necessary information as they are lowered down into the earth. This natural radiation detector is often followed by another device, which irradiates the earth with neutrons. The device **measures how many neutrons** are reflected back; a high neutron reading indicates the presence of hydrogenous material such as oil or water in the surrounding rock. Then other radiation detectors follow which **record new radioisotopes that have been created** by the brief neutron irradiation, and thus tell what other elements are present.

Consumer Products

Several characteristics of radioisotopes make them useful in many consumer products. The radiation they emit ionizes the air, a process that eliminates static and facilitates the flow of electricity. Therefore, they are used as **static eliminators** in copying machines and printing presses, to keep the papers from sticking to each other, in **lint brushes** to eliminate static electricity in clothing, and in **shrink wrap packaging** used on everything from vegetables and soft drinks to retail hardware items. An isotope of americium, an element not known before the nuclear age, is used in millions of inexpensive home **smoke detectors** installed in 88 percent of U.S. homes. These detectors are said to increase the chances of surviving a fire by fifty percent.

Many consumer products are **sterilized** to be made free of germs and allergenic irritants; these include cosmetics, bandages and many products for baby care. For heat-sensitive items such as medicines, plastics and ointments, radiation is the only feasible sterilizing process. **Reflective traffic signs and exit signs** are made easier to read at night by incorporating radioactive materials such as tritium.

Research

The great majority of Nobel Prize winners in medicine relied on radioactive materials in their research. Here are a few examples of other research applications.

The natural radioactivity of carbon enables us to **estimate the age** of prehistoric and early historic artifacts. The natural radioactivity of uranium works over a longer period and is used in **dating geological eons**. Measuring the **relative abundance of various stable isotopes**, using techniques developed in the nuclear program, gives information as to the age and composition of meteorites and the stability of the arctic ice caps.

By **neutron activation analysis** scientists can measure extremely small quantities of impurities in substances, a capability that enables researchers to measure engine wear, helps museums to expose fraudulent antiques, and police to link suspects to crimes. **Low-energy x ray examination** helps artists restore paintings damaged by age or by poor restoration attempts. Use of **radioactive tracers** has become a major tool for following the path of complex chemical or biological processes.

Electric generating units, powered by small reactors or by radioisotopes, have powered 25 missions into space, and other units have been transmitting information from unmanned satellites and from the surface of the moon. Similar nuclear generators power **weather-monitoring stations** in Antarctica and other remote locations on earth.

Agriculture

A quarter of all the world's harvested food is lost to spoilage, decay and insects. Nuclear technology is fighting this problem several ways. **Sterilizing thousands of male insects** in the laboratory with high doses of radiation and then releasing them into an infested area has proved to be an effective way to reduce crop-threatening insect populations without using environmentally harmful pesticides. **Tagging fertilizers with radioactive tracers** enables agricultural researchers to measure how fertilizers are taken up into various plants. On the basis of this information, farmers are advised as to the best time to apply fertilizer and how much to use for optimum results, saving money and minimizing environmental impact of fertilizer run-off.

The radioactive carbon isotope C-14 is continuously created in the atmosphere from cosmic rays interacting with nitrogen in the air. Living things and surface water keep exchanging molecules with the air, and so they too maintain about the same concentration of C-14. Groundwater can be tested for C-14; if the C-14 concentration is low, that means the water has not been in contact with the atmosphere for a long time. Such underground pools of water should not be pumped up for agricultural use, or they will soon run dry.

Irradiation can destroy noxious bacteria, viruses and molds. **Irradiated food** can be stored in sealed bags at room temperature for long periods without spoiling. Properly done, **irradiation can kill salmonella, trichinosis and other harmful pests** and reduce the need for chemical pesticides and antibiotics. Sterilized food is used widely in more than 25 countries and by astronauts and military personnel where refrigeration is a problem. It is only recently being introduced for widespread public use in the U.S. for spices and some fruits and vegetables, and has now been approved for many types of meats.

The atom has impacted nearly every corner of life in the industrialized world. Not just indirectly through the political and economic effects of nuclear bombs and power plants. But up close and personal in nearly every product we use. That is a factor that must be considered in every decision we make regarding the future of nuclear technologies.

The Washington Post — FINAL

THURSDAY, JANUARY 20, 1994

Record Cold Threatens Power Supply; Businesses, Governments Close Here

Deep Freeze

The Washington Post — FINAL

FRIDAY, JANUARY 21, 1994

Power Crisis Ebbs as Many Stay Home

State of Emergency in D.C. Is Lifted; Forecast Calls for Above-Freezing Weekend

Unlocking the Potomac for Pepco

Coast Guard Cutter Leads Barges Through Ice to Get Oil to Power Plants

By David Montgomery

The frozen Potomac River stretched ahead like an eerie white ribbon in the moonlight shortly after midnight yesterday as a big Coast Guard cutter began carving a channel for two precious charges trailing behind.

The cutter's mission: escort two oil barges from the Chesapeake Bay to the District and ease the local power ...

... Gentian. "But when they said they weren't going to have any lights in Washington, we told them we'd take more risks than normal."

In the cutter's wake, private tugboats pushed the barges, which carried 60,000 barrels of oil bound for Potomac Electric Power Co. generating plants in the Washington area. Pepco had asked for the Coast Guard's assistance.

At the other end of the 16-hour ...

"That oil is going to allow me to keep [the power plants] functioning and operational," said Felton, vice president for fuels at Pepco.

The barges could never have made the 90-mile trip from a petroleum terminal in St. Mary's County, Md., without the cutter's reinforced steel hull to run interference through ice that was a foot thick in places. If they had tried, the ice could have punched holes in the hulls and created an oil ...

Related Stories, Data on Closings

Figure 13.1 Power crisis proves the importance of having reliable electricity at all times (*The Washington Post*)

13. Bulldozing the Garden of Eden
AND OTHER CONCLUDING THOUGHTS

The Cultural Chasm

It's not surprising that persons who are uneducated or uninformed might fear and suspect new technologies they can't understand. But I find it strange and very sad that some people who call themselves intellectuals, who delight in activities of the mind, respond to technology in an almost anti-intellectual way. They announce defiantly that they cannot understand anything technological and have no wish to do so. They act as if they consider it vulgar, not worthy of their attention.

In the late 1950s, the British scientist/writer C. P. Snow started lecturing and writing in stark terms about this situation. In his book *The Two Cultures*, he laments:

> "Between the two a gulf of mutual incomprehension—sometimes (particularly among the young) hostility and dislike, but most of all lack of understanding. They have a curious distorted image of each other. Their attitudes are so different that, even on the level of emotion, they can't find much common ground....It is all destructive....This polarisation is sheer loss to us all."

Snow defined the adversaries as the scientific and the literary communities, but it is clear that the conflict included, on the one side, all those whose profession involves the material world against those who call themselves humanists.

Looking back from nearly fifty years later, with my own experiences fresh in mind, I see the problem as having many facets and more than one source. A number of forces are at work. There is the medieval tradition that those whose life revolves around reading and writing are the intellectuals. Such a person might concede that another fellow (but probably not a woman) was cleverer than he, but this would not make that fellow an intellectual. Snow reports that eminent scientists of his day were puzzled that the literary elite "while no one was looking took to referring to themselves as 'intellectuals' as though there were no others." And this, I believe, accounts for some of the estrangement.

In addition, I have learned that many people view technology as an abstract, mysterious, alien force, proceeding on its own destructive mission, uncontrollable by mere mortals. This view is explicitly set forth in gloomy pronouncements by such writers as Jacques Ellul, Lewis Mumford, René Dubos, Theodore Roszak and Charles Reich, warning that technology, aided by advertising and other evil capitalist devices, "turns out what it pleases and forces people to buy." This picture is bolstered by the notion that, since people insist on buying things of which these writers disapprove (such as fancy automobiles, washing machines and vacations at exotic locales), they must be doing so against their will. Despite the eloquent arguments of such writers praising the advantages of rural over urban life, people of almost every culture have flocked from the farms to the cities since the dawn of civilization, and are still doing so at an increasing pace.

While there are important benefits that accrue from a simpler life and less dependence on mechanical and chemical support, it is not the fault of technology that many people choose comfort, convenience and instant gratification. As the perceptive engineer-sociologist Samuel Florman has demonstrated, what these anti-technological writers are asking for is nothing less than a change in human nature. There is some merit in their argument, but their focus on technology as a source of the problem does not stand up. C. P. Snow wrote bluntly:

> "Intellectuals, in particular literary intellectuals, are natural Luddites....If the scientists have the future in their bones, then the traditional culture responds by wishing the future did not exist."

I think we must accept the fact that technology is a natural activity that some people like to develop, the way beavers build dams and bees make hives and spiders spin beautiful webs. Those who take up engineering activities are not all "Dr. Strangeloves," but encompass the entire spectrum of character and personality. In the earliest days of our prehistory, some of us naturally chose to hunt for meat, others gathered edible plants and still others built shelters, pottery and crude boats. These were all useful activities, proudly carried out and warmly appreciated. Most technologists, like most of the rest of us, love their children, are liked by their neighbors, and want their work to make life better for all of us.

The Existential Pleasures of Engineering

Engineering is something people enjoy, like fishing or writing music. Samuel Florman wrote about it in his wonderful book, *The Existential Pleasures of Engineering*. He quoted the loving language used in the Old Testament to describe the craftsmanship that went into building the temples, and the ecstasy inherent in the building of ships, as described by Homer and others through the ages:

> Although the *Iliad* and the *Odyssey* are concerned mainly with the affairs of noble warriors, Homer makes clear repeatedly his admiration for the craftsmen of his age. He refers to "an expert carpenter, who by Athene's inspiration is well-versed in all his craft's subtlety." The manufacture of a shield is credited to Tychios, "far the best of all workers in leather." We are told of a famous smith, "who understood how to make with his hands all intricate things, since above all Pallas Athene had loved him." There is an architect fitting roof-beams, a bowyer making a bow, a swineherd who has built a handsome pigsty. Armor is wrought "carefully" or "with much toil." A doorway has been "expertly planed…"

We emerge from the world of Homer drunk with the feel of metals, woods and fabrics, euphoric with the sense of objects designed, manufactured, used, given, admired and savored. If this be materialism, then our ideas about materialism seem to be in need of revision.

These are natural healthy human reactions to creative constructive efforts that please the senses and serve a physical need. The ancients were close to their handiwork and were aware of, and understood, the human effort that created it. But one can argue that we moderns seldom see the craftsman at work, and in fact most of our artifacts are machine-made elsewhere by processes we don't even know. Is any of this applicable to us? Of course it is. We can still take pleasure in a well-made house or car, and the more we know about the technology that creates our artifacts, the more pleasure we can get from them. Florman concludes that reading stories of earlier days "reinforces our intuitive belief that engineering is a basic instinct in man, the expression of which is existentially fulfilling." To which I say *Amen!*

To Engineer Is Human

Henry Petroski is another engineer whose books insightfully describe his profession in homely terms anyone can relate to. In *To Engineer Is Human* he compares the engineer's task of creating a new design to the task that faces anyone planning to take a trip, say from Chicago to New York. He examines the first decision—how to get there—and notes that although one *could* go by bicycle, by hot-air balloon, or even by boat via the Great Lakes, most people would not give those options much consideration except under special circumstances. That leaves train, plane, bus or personal car. He then explores the advantages and disadvantages of each, noting that a single traveler watching his pennies might choose the bus, whereas a large family might drive their car. Others, stressing comfort and convenience might fly; still others might enjoy the train ride.

There is no universally correct solution to the question: How should one travel from Chicago to New York? He carries the analogy further to questions like: Should they stay in a low-cost motel in New Jersey and travel back and forth to New York each day, or should they stay in a higher priced, conveniently located hotel in Manhattan? "Clearly the choices can appear endless," he notes.

Then he comes to the heart of the matter:

> Engineering design is not much different. Many objects of design are no more exotic than spending two weeks in New York. Even if you and your own family have not done it before, there are plenty of people willing to give you advice about what to do and what to avoid. There are books on the general subject of New York and others on such specialized aspects as the city's museums, restaurants and shopping opportunities....And the availability and price of hotel rooms, theater tickets, and restaurants can be obtained over the telephone. In short, there is a wealth of experience and information out there for the asking.

He adds, "The engineer designing a new highway bridge also has a wealth of experience available to him." He goes on to describe how the engineer will seek out other engineers who have built similar structures, will read books both general and specific, and will study various design manuals, engineering codes and standards, materials and construction reports, articles in technical journals, and all the other relevant information he can glean. This is a homely process, not at all mysterious—one we can all understand. The fact that the engineer calculates structural strength and rigidity using complex mathematical formulas and computer programs unknown to the lay person should not be any more intimidating than the process by which a professional golfer chooses the right club for a difficult shot, or the know-how by which an artist such as my wife Mary selects paints from her palette to make just the right color for her sunset.

The rest of us know those particular skills are far beyond our reach, yet the basic process of playing a golf tournament or painting a picture is something we can all understand. I can pick up a pencil or a crayon and draw a picture for my granddaughter. Although neither she nor anyone else can tell whether my picture is intended to portray a horse or a dog (although I always make my horses *bigger* than my dogs), the activity of painting does not strike me as alien or unnatural. The Olympic gymnast's perfect maneuver may leave us breathless with admiration and wonder, but we could picture ourselves doing it (in our dreams!).

Some professions are not like that. Few people can imagine what it must be like to spend all day working on theoretical physics or composing a symphony. These are arenas where even extraordinary skills do not suffice; the very fundamentals of the process seem to be beyond the comprehension of ordinary mortals. But my concern here is not with theoretical physics or music composition. I would just like to show you that engineering is much like things you already know how to do. Like planning a trip or other complex tasks, these things involve many decisions, based on uncertain and incomplete information, which in important ways is not quite like anything anyone else has ever done before. No one else has ever tried to take *your* family to this particular place under these particular circumstances. Some of the decisions are difficult, close calls, and even if you think you've made the right decision at the time, you may look back and say, *I'll never do that again!* Welcome to the club! What you've done is not that different from the job of the engineer.

The Myth of the Riskless Society

Realizing that engineers are but fallible human beings, we might ask that they stop taking chances and stick to what has already been proven satisfactory. Why should they keep putting us at risk? Petroski answers that one, too:

> Every new bridge could be an exact copy of one that has already stood the test of time, but traffic on the new bridge could never be allowed to surpass that on the old. No new materials could be used, and no new bridge could be located on a river that did not possess the exact foundation and wind conditions of existing successful bridges… we would in effect not allow any bridge to be built where one had not been built successfully before. For no place is quite like any other, no traffic pattern like any other.

You may find yourself thinking, *Come, come, now, we don't have to be that picky. Just don't make any* important *changes.* But then I must remind you that 114 persons were killed, 200 injured, and $3 billion in lawsuits were filed when the Hyatt Regency walkways collapsed in Kansas City on July 17, 1981, all because the construction contractor decided to place a machine nut under the second balcony instead of under the first, where the designer had shown it. And the famous American

Airlines DC-10 crash in Chicago in 1979 was caused by "excessive and unanticipated abuse" during maintenance of a flange on the engine which is part of the assembly that mounts the engine onto the wing. Any such detail becomes important when it leads to failure. Seldom does a design fail *in concept*. "For want of a nail ... the kingdom was lost."

There are two more points that should be made clear. Although failure of a structure gives unequivocal proof that its design, construction or operation is in some way inadequate, continued successful operation is not proof that the structure is satisfactory. A dramatic example of this was the celebration of the fiftieth anniversary of the Golden Gate Bridge in San Francisco. Here is a structure that had performed flawlessly for half a century, including a celebration shortly after its opening, when 200,000 people walked happily across its 4,200-foot open span. Yet fifty years later, when the "open bridge" celebration was repeated, the roadway and sidewalks were so packed with 800,000 people that no one could move for more than an hour, creating a loading on the bridge far in excess of the normal, or even heavy, traffic.

During this period, observers noticed that the bridge began to sway several feet to each side, and some of the hanger cables became slack, which they are not supposed to do. The arch in the middle of the span flattened by ten feet—another condition not anticipated by the designers. Luckily, the people were moved off the bridge in due course and no failure occurred, so the entire affair was a "non-event." But there are other structures that have collapsed after decades of "successful" operation, due to changed traffic or wind conditions, or undetected corrosion or crack propagation.

The other option for a wary public is to say, *OK, engineers are fallible, and we can't avoid some risk. But we'd like to set a limit. Can we limit the risk to ten deaths? Or maybe a hundred?* This sounds like a reasonable position, but how do we do it? The highway leading from the stadium in San Francisco was crowded with cars returning from the 1989 World Series game. If that highway had collapsed, thousands might well have died. It did, in fact, collapse a few hours later when the earthquake struck, but only about fifty people were on it. Should that risk be considered a fifty-person risk, or a risk to thousands? When a major airplane crashes, two or three hundred people may be killed. But, as we've seen, if it crashes into a crowded building and explodes, the dead could number in the thousands. The failure of some dams could threaten hundreds of thousands. If the rusty tanks storing chlorine for a city's water treatment were to fail, the entire city could be in mortal danger. How far do we speculate?

Technological Man vs. Mother Nature

Much of the deep-seated fear of technology rises from the belief that we are defiling Nature, that we are doing something unnatural, perverted, evil. We are bulldozing the Garden of Eden. Leaving aside that the longing to return to Eden is itself theologically heretical, let us try to understand what it really means to call some of our activities "natural." Sometimes it seems to mean merely that one uses only technology at least a generation old. For example, the Amish (or "Pennsylvania Dutch") people, who cleared virgin forests to plant hybrid corn, are considered "natural" people because they don't use tractors or chemical fertilizer.

Technology makes it easier to convert eroded wasteland and arid deserts to cropland. Should we not create gardens in the desert if it requires deep-well technology to do so? Is self-realization abetted by poor sanitation, high infant mortality, illiteracy, leprosy and grueling workdays that allow no leisure? We are delighted to find that some birds and some primates use tools of their own devising. Are we to be denied that right? Should we limit our dams to the size of a beaver's? In our newly found concern for the Earth, the Species, and Life, let us not overlook individual human persons with their legitimate and, yes, natural dreams for a better life for themselves and their loved ones.

"Out of Control"

One issue that technologists have not yet come to grips with, because it is just now beginning to be understood, is the consequences of trying to impose controls on any highly complex, interactive, evolving system. We see this clearly with the economy and we are beginning to see it with the environment. Kevin Kelly, in his astonishing book *Out of*

Control, talks in detail about this problem. He shows that many kinds of complex, interactive systems want to evolve in their own way. This they do with remarkable efficiency, but we may not like where they are headed. If an "uncontrolled" economy wants to produce a handful of rich people and starve the masses, or a natural ecology wants cockroaches to replace humans, we decide to step in. We regulate the interest rates and tax structures, and we kill off the cockroaches, rats, and crabgrass. This provides a quick fix that may last until the next election. But we find that the system seldom wants to work with us to achieve the result we want. This is a problem for which no one has yet come up with a workable solution. I suspect the answer will prove to be unlike anything we've tried so far.

What Are We Fighting Over?

Despite all the energy that goes into dividing people into Us *vs.* Them, you can't help noticing that virtually everyone wants a world free of poverty, injustice, war and pollution. We all recognize the satire of Tom Lehrer's new-age folksong: "I'm against poverty, war and injustice, unlike the rest of you squares." So what is it we are fighting about?

Some people believe that science and technology offer the only effective tools for understanding the nature of the world and for achieving these goals. Others believe that our focus on technology has blinded us to spiritual truths that are the ultimate reality. But both sides have similar values that they'd like to see incorporated into a new world. I am convinced that, despite the continuing advances of science, the human craving to satisfy spiritual and artistic needs will not diminish. I also believe that most of us will wish to take advantage of the opportunities for greater comfort and fulfillment offered by technology. So I don't see either side vanquishing the other.

Second, I've found that it's lingo, the arcane jargon of the technologist, that convinces many people that the subject itself is mysterious, alien and beyond their reach. But this problem exists with any specialized human endeavor. Listen to teenagers taking about music or working on their computer upgrades. Webster's Second International

Dictionary lists 64 different terms for the sails and parts of sails on an old-fashioned square-rigger. Eskimos are said to have 27 words for snow. And some preteens know dozens of long Greco-Latin names for dinosaurs. Just because you may not know many dinosaurs by name does not mean that the preteens are privy to a field of arcane knowledge beyond your ability to comprehend. Lingo may hide knowledge from you, but you should recognize it as a flimsy barrier, easily breached.

Further, you need not fully understand all nuances of a subject or be skilled in its techniques to have a comfortable feeling for the subject itself and be able to make common-sense judgments about it. People with no professional medical expertise do not hesitate to recommend treatments for minor health problems.

There Are Some Concepts You Must Learn

In addition to the lingo—a few phrases and terms of art—there are also some concepts that must be understood when trying to evaluate a new field of knowledge. Therefore, at various points throughout this book, I've brought in some background information, to give you, one spoonful at a time, some of the tools and ideas you'll need, to understand scientists or engineers when they try to explain what they do.

This was not to try to teach you quantum mechanics or thermodynamics—that's the *content* of science and engineering. I just wanted to explain the *language* and *concepts* that scientists and engineers use. That's a lot easier. But it's even more important than content if there is to be understanding. If a scientist is trying to explain what he does and he's speaking French, it wouldn't do you any good to know quantum mechanics and thermodynamics if you didn't know French.

I hope you found that it's a lot easier to learn the language of science and engineering than it is to learn French. The ideas are simple and straightforward and delightfully explicit. Because the product that scientists and engineers produce must be precise and unambiguous, the language they use when they talk about their work must also be clear and precise. This should make it easier to see through some of the traps and pitfalls clothed in the language of science that advertisers and political advocates often set for the unwary.

You *Can* Understand What You Need To Understand

Many people are convinced that they have no ability to understand anything scientific or technological, and, of course, such a deeply held belief can be self-fulfilling. My son, Teed, a philosopher/musician in San Francisco, once had that view. Here's a man with an original, brilliant and open mind, who publishes highly technical essays in peer-reviewed philosophical journals. But he felt that technical matters involving machinery were forever and inexorably beyond his grasp. One day he told me that his Volkswagen had broken down and wouldn't operate.

"I was out in the country, alone, so I got out and opened the hood," he told me. "I had no idea what I'd find there, or what I could do about it, but what else can you do?" I agreed that his logic to that point was unassailable.

"I saw a little doo-hinkus with a hook on it, hanging loose, that looked as if it should be connected to something," he continued. "The only thing it could reach was a thing that had a little loop which looked as if it were just made to have a hook through it. I pulled on the thing with the loop and found that it made the accelerator pedal wiggle. The other piece moved things that seemed to end up in a round gadget that could well have been the carburetor. I hooked them together and found that the car was working again. I had repaired a car!"

This had a real impact on his thinking. He had no illusions that he now understood automotive engineering or that he was a qualified car mechanic. But he now knew from direct experience that he *could* understand some mechanical details if he merely applied the same sort of mental effort that he would apply to any new idea in his own areas of expertise.

A few years later, he had a similar experience in another field. He and his wife Diana sponsor a series of concerts each Christmas season, and one year they hired someone to handle the promotional aspects. They were impressed with her promise to get grants to help finance the effort, and her ability to get newspaper publicity and radio time. "We could never do that," they said. "She is an invaluable asset to us. We wouldn't begin to know how she goes about it." The following year, they suddenly found themselves without their publicist and with very little time to prepare for the next concert series, and fighting a lawsuit at the same time.

They had little choice but to "wing it." In a matter of weeks they found that promotional activities, like every other human enterprise, can be learned and mastered, and they wondered why it had previously seemed so intimidating. In addition, they were able to get financial backing from various sources on their own. Since they could afford only the barest amount of legal advice, Teed took to learning enough legal lore to mount a persuasive and effective legal defense on his own, and he won his case. Again, the lesson was profound: "I'm convinced that any intelligent person can learn enough in any field to carry on a meaningful dialogue, and make judgments and decisions," he concluded. "You don't have to be a law school graduate to discuss alternatives with a lawyer intelligently and to direct your own case with confidence. And you don't have to be a professional mechanic before you open the hood of your car and maybe even fix it." He and I are convinced that this is a part of the answer to the intercultural wars—but only a part.

Even after we penetrate the battle rhetoric to where we can hear the real voices of the others, we still have a problem of understanding. *What do those words mean? I don't understand that concept. What is the significance of those numbers?* As we've seen, in addition to terminology there are often basic facts and concepts that are necessary to grasp before you can go on to understand the real substance of a topic. *What is half-life? How much natural radiation do we get? How big is* 10^{-3} *compared to* 10^{3}*? What's the difference between an average, a mean, and a median?* Sometimes you can't even understand the meaning of a simple sentence if you don't understand concepts such as these. But once these concepts are understood, the other aspects of the subject may then become clear. There is a great deal of wisdom in the old folk saying, "It's not ignorance that causes the problems, so much as people knowing things that ain't so." In this situation, it is important to extricate and clarify the factual or conceptual nugget, and separate it from the political or sociological question in which it is embedded. Then, with the facts clear and agreed on, meaningful dialogue on the subject is possible. A word like "energy" means one thing to a spiritual healer and quite another to a technologist. We have to define how we are using it in a given circumstance.

It's a little like learning to swim or ride a bicycle. If an adult who never learned to swim is thrown into the water, he's sure he'll drown—and he may. His eyes, ears, nose and even lungs may feel the water and protest frantically. He will be sure that no air-breathing animal could possibly survive in such a hostile environment. He will swear that if he survives this ordeal he will never again venture into the water. Yet, with a little guided effort and some friendly help he can learn to splash around happily with little effort and no fear. So it is with the world of science and technology. If you've never tried it, you may be surprised to find how easy it is to become comfortable with it. That doesn't qualify you for the Nobel Prize, but it does enable you to feel at home in a new environment and share the experience with others.

The Tannen Effect

There is another, quite different, barrier to communication and understanding that I call the Tannen Effect, after Deborah Tannen, the noted professor of linguistics at Georgetown University. The type of situation that Professor Tannen made famous involves a wife who has suffered an indignity during the day that hurt her deeply. When her husband comes home, she starts to tell him about it, looking to get his understanding, sympathy and succor. Her husband listens intently, for he loves his wife and can see that she is distressed. But he also prides himself on his quickness of mind, and after two or three sentences from his wife he is confident he understands the situation. He cuts her off, assuring her he fully understands and intends to do something about it. He will call the man who treated her so rudely, explain to him how deeply he hurt her, tell him that his action spoiled her plans for the afternoon (which the fellow could not have known about), and straighten the whole thing out. *No, no, he's glad to do it. What's for dinner?*

He can't understand why she runs off crying, and she can't understand why he is so cold and unfeeling. "She doesn't want to solve her problems; she just wants to whine about them," he complains later to a friend. Her view of the situation is quite different: "He wants to trivialize everything that happens to me, to brush it off with some quick-fix. He doesn't really care how I feel."

Superficially this situation is not at all like the hostility and misunderstanding that exists between technologists and their adversaries. The husband and the wife profess love for each other and a desire to understand. There is virtually no "lingo problem;" neither party is using terminology unfamiliar to the other. But despite the differences between the two situations there is an important lesson here nonetheless. It is simply this: Even though each party would be confident that they fully understood the words and the meanings conveyed by the other, yet they misunderstood the basic purpose for which each entered into communication. The wife wanted to explain how she felt and thereby get understanding, sympathy, and thus to deepen the bond between them. The husband wanted to solve a problem and demonstrate his love thereby. Their intents were at odds though each thought they were working toward the same objective. The same is often true of technologists and their adversaries.

Tannen makes clear this is not just a man/woman thing. Each of us runs into this problem from time to time. I tell a story, just because I think it's a story the listener will enjoy. But I find myself interrupted repeatedly with what I consider completely irrelevant comments such as, "Well, I can see why he did that," or "You certainly can't blame her for getting upset." My purpose was completely narrative, but the listener assumed I was trying to convey a judgment and felt compelled to support or contest that judgment. Or, in another instance, I had an awesome experience in a walk through the country on one of the first days of spring. I am trying to describe the colors, the brightness, and the feel of the newly warm breeze on my face, when the listener interrupts to tell me that there is a much better general store in the next town. He thinks my purpose is to relate facts, when my intention is to convey an emotional experience.

The relevance here is that we cannot communicate properly if we misunderstand the purpose behind the communication. That purpose cannot always be ascertained from the words being spoken. When we hear a spokesman for the Government Accountability Project raise questions about plant safety, we should recall that his organization is on record saying, "Let's face it. We don't want safe nuclear power plants. We want *no* nuclear power plants." And when energy critic Amory Lovins raises questions about nuclear power, we should not assume he shares our objective of providing clean, cheap, abundant energy since he has written, "It would be little short of disastrous for us to discover a source of clean, cheap, abundant energy, because of what we might do with it."

Part of my purpose in this book is to show people that they need not fear technology blindly; that if they try, they can understand enough to appreciate what is good and oppose what is bad. In particular, I want to show that nuclear power is an understandable and beneficial technology. I recognize that some people will disagree with me on this judgment, but I hope they may still benefit from the information.

The Myth of the Technological Imperative

At the root of many people's fear of technology is the Myth of the Technological Imperative, a belief that anything that has become technologically possible will inevitably be introduced into the world, regardless of how harmful it may be—nothing can stop it. Behind the perception of a Technological Imperative lies the notion of a Technological Elite. But this, too, is a myth. Important technical decisions in the industrialized world are not made by engineers and scientists, but by politicians, financiers, lawyers and businessmen. These persons may be technically illiterate, but they make the decisions. And they are subject, to some degree, to pressure from the ballet-box, the stockholders or the marketplace.

For example, after World War II, public concern over sickness and deaths from food poisoning led to the development of a variety of preservatives. Crop losses from insects led to DDT. Then, concern over side-effects of these materials created a rash of "no preservatives" foods and a ban on DDT. (The dramatic drop in malaria deaths in some tropical regions when DDT was first used, and the appalling resurgence of deaths when it was outlawed, have initiated a third round of evaluations in those countries. In Sri Lanka, DDT cut the malaria cases from 2 million to 17 (yes, just 17!), and the deaths from 12,000 per year to zero. But when DDT was banned, the malaria cases rose to 3,000 in 1967, 1 million in 1968, and 2½ million in 1969.)

This does not prove that we should now start using DDT again. The point is that just determining that a given technological solution may have undesirable side-effects is not enough. There may be other side-effects—unexpected and serious—from *not* using a given technology. The power to decide rests in The People, through their fallible representatives, not in an abstraction called Technology, nor an Elite called Technologists. And the decision may be a bum one. There are not many easy answers.

Fear of Scarcity

Just before we stripped the last of Europe's forests for firewood, we discovered coal. And just before we turned England's atmosphere into an uninhabitable smog, we discovered oil. Now, as we realize that oil is more valuable as a base for the whole petrochemical industry, for plastics, medicines and other specialties, and as fuel for airplanes, we can stop burning it in the huge quantities necessary to make electricity and heat buildings, and use nuclear power for electricity and sunshine for heat.

Present commercial reactors get most of their energy from fissioning the rare isotope uranium-235. It is estimated that if we made all of our electricity this way, we might run out of U-235 in a few hundred years. But we long ago demonstrated that reactors can "breed" their own fuel. By proper design, the abundant isotope uranium-238 can be converted into fissionable plutonium while the reactor operates. An alternative breeder design converts non-fissionable thorium into fissionable uranium-233. In both cases, we have now demonstrated that we can actually operate so as to produce more fuel than we use up! This would extend our nuclear fuel supply for thousands of years into the future. In fact, the energy content of the uranium now already stored at

American uranium enrichment facilities could produce energy equivalent to over three times all the world's known oil resources plus the energy of all the recoverable coal resources in the United States.

Until breeder reactors are in widespread use, there is another way to conserve. Current nuclear power plants have initial license permits for 30 or 40 years, and these licenses are approaching expiration for most of our plants. Unless the plants have been damaged in some way by excessive corrosion or neutron embrittlement, there is no reason that they should not have their licenses extended for another 30 or 40 years. These plants have proved themselves in years of reliable operation, and their initial costs have been paid off. However, when consideration of license extension first arose, many politically-minded Public Utility Commissions (PUCs) used the occasion to promote their anti-nuclear agenda, dreaming up grotesque financial scenarios to prove it would be cheaper to scrap the plants and build new ones!

They hypothesized, despite evidence to the contrary, that the nuclear plant would be shut down a large part of the time and that maintenance costs would be high, and they assumed that natural gas would remain cheap for decades into the indefinite future. A number of nuclear plants were shut down on this irrational basis, and more were scheduled. An unparalleled opportunity to use these proven, paid-for plants was being cynically sacrificed for political ideology. Particularly with the abundant availability of already-processed, clean fuel from nuclear weapons, these plants should be considered an unprecedented natural resource. Fortunately, the Bush administration, aided by increasing gas prices and increased pressure to reduce production of "global warming gases" is encouraging reopening the nuclear option.

We are learning to conserve other resources. We now use tiny silicon chips in satellites to replace tons of precious copper in transatlantic cables. Silicon is second only to oxygen in abundance throughout the world. Aluminum, also in virtually unlimited supply in the earth's crust, is being substituted for copper in many large electrical systems. This makes electricity cheaper, and cheap electricity is the key to producing aluminum in large quantities. It is by such means that we have created decreasing scarcities.

The assumption that conserving energy should be an overriding goal in itself has led needlessly to some undesirable ends. For example, we are told to seal up our houses and offices tightly, to conserve energy. Then we find people getting sick from paint thinners, cooking fumes, plasticizers and other pollutants from within the building. Elaborate schemes have been proposed to deal with these pollutants, but people are beginning to realize that the simplest and most appropriate solution may be just to pull a moderate amount of fresh air though the building, even if this adds slightly to the heating or air-conditioning bill.

I witnessed a dramatic display of the havoc that can be caused by minimizing electric power reserves. For several days in January 1994, the front page of *The Washington Post* was covered with headlines and stories of the electric power crisis that a few days of cold weather created in the nation's capital (see Figure 13.1). A U.S. Coast Guard cutter was called in to break ice in the river and lead oil barges to the power company. "We'd rather not go there at night," said the cutter's skipper. "But when they said they weren't going to have any lights in Washington, we told them we'd take more risks than normal." But even so, the U.S. Government and most local businesses were shut down until the crisis passed. The local power company ran a full-page ad after the crisis:

> ... [the cold] put remarkable strains on power plants throughout the mid-Atlantic ... fuel availability was low. Natural gas was unobtainable for generating electricity, and oil deliveries were hampered by iced-in barges and hazardous roadways.
>
> These extreme conditions forced Pepco and other utilities to take radical steps to drastically reduce energy use. Otherwise we faced the risk of widespread, uncontrollable outages. The situation became so critical that we had to request business and government to close.
>
> Thankfully, disaster was averted.

All this could have been avoided if the power plants had been running on nuclear fuel, which can run at full power for a year or more between refuelings. It could also have been avoided by resisting the misguided shouts to keep trimming our "wasteful" energy reserves until there just wasn't enough.

Restricting Energy Use as a Moral Issue

The argument for restricting energy use often takes on a moral, quasi-religious tone. No one is in favor of waste. That is not an issue. But let us examine such slogans as "With only 4 percent of the world's population, the U.S. accounts for a third of the world's energy consumption," or "The rich should live simply so that the poor may simply live." The facts are by using energy in a complex society, the U.S. produces one-third of the world's goods, including food and fuel shipped to other countries. The average American farmer produces enough food to feed 137 people. What better use could be made of energy? What else should we be saving it for?

A field of grass is more energy-efficient than the cow that grazes on it. And the cow in turn is more energy-efficient than the tiger (or the human) that eats the cow. Does this imply some sort of moral hierarchy? We are told that we should pursue certain activities such as growing corn for alcohol rather than building other types of power plants, because growing corn is "labor-intensive." That term merely means that it requires lots of human labor. But of course, trying to do things with *less* human labor has characterized most of human inventiveness for the last five thousand years. We should not try to reverse that trend without some very good reasons.

The Importance of Energy in Our Lives

The risks associated with not having energy available when needed are difficult to envision, but very real nonetheless. When power failure hits a police station or a hospital emergency room, or heat and lights go out in a large housing project, then the importance of energy for maintaining a safe and healthy existence can become vividly obvious.

Energy is the critical ingredient for solving most of our material problems. For example, we hear of dangerous shortages of water in various locations, but three-quarters of the earth's surface is covered with water, much of it a mile or more deep. Use of energy can turn it into pure drinking water, and more energy can pump it to wherever it is needed. We hear of impending shortages of raw materials, but this too is easily solved by the use of energy. We used to get many minerals, from iron to uranium, in ores that had over 50 percent of the desired element. Now we can economically recover minerals from ores with a fraction of one percent. In the extreme case, we can always go to the oceans, which contain nearly all the earth's minerals, albeit in dilute form.

Must we really choose between technology and spirituality? Of course not. One of the most hopeful signs for the future is the depolarization of some of the hostility of the '60s. We find long-haired capitalists with an earring, folk concerts using high-tech acoustic equipment, and computer networking dream interpretation workshops. Culture commentator John Naisbitt points out that we instinctively pair up "high-tech" with "high-touch", that we balance the two against each other, getting the best of each and reveling in the difference—like sweet and sour pork or a baked Alaska. He cites hospitals vying for the latest digital diagnostic equipment while sponsoring or working with hospices, birthing centers and home care programs.

Real-World Experience

There are still some academics who write in their air-conditioned, well-lighted offices, that people in developing countries should keep to a primitive, "natural" lifestyle, guiding a wooden plow behind a tired ox and singing their happy native folksongs. But real people in those lands aspire to freedom from drudgery and a chance for what they perceive as a better life, for the same reasons that the rest of us have those aspirations. This does not justify our exporting some of the worst aspects of our culture to these people, such as drugs and greedy CEOs, but neither should we presume to decide for them that they should not want the things that abundant energy makes possible.

We think of "the glory that was Greece and the splendor that was Rome," and we ask, *Can't we follow that path? They did it without modern technology*. True, but they paid a price we are no longer willing to pay—slavery. Thirty-four thousand free men and women in classical Greece depended for their necessities, luxuries, and leisure time on the grinding toil of 300,000 slaves. Ancient Rome required about the same ratio: 15 to 20 million citizens supported on the sweating backs of 130 million slaves. This ratio, nine slaves per person or 36 slaves for a family of four, is not far from the energy used to support a citizen in the modern industrialized world. If we estimate the continuous output of a healthy, hard-working human slave at 40 watts, we find the typical family in ancient Rome or Greece consuming about one-and-one-half kilowatts of human energy, plus at least that much animal "horsepower." Today we can provide that same energy in a much more efficient form, electricity. No human slaves are required, and no animals need be chained to treadmills.

As we look at various cultures around the world, we see a close correlation between increased use of energy and an increased longevity, higher literacy, lower disease incidence, and other indications of a materially better life. Hospitals, schools, libraries, good housing, transportation and sanitation systems, public health programs—all these things require expenditure of energy. There is much we can learn from primitive people, and there are many aspects of their approach to life that are superior to ours. But both experience and common sense tell us that neither spirituality nor wisdom need be impaired by the introduction of the means of providing for decent material well-being for all the people of the world, including our own.

We cannot eliminate the effect of our presence in the world or ignore it. All we can do is observe as accurately as our instruments and our brains permit and act as seems appropriate. Our power gives us great responsibility, but we have a right to be here. Indeed, we can be nowhere else. We are told that things are scarce because Technology is "using up" Nature. The crucial point here is that a distinction is made at all between Technology and Nature. Once you have the idea that a farm is natural and a factory is not, you have set a whole way of looking at how we use tools, and which of our gifts we may enjoy, and whether it is appropriate for primitive people to learn to drive jeeps. Such questions open up trails to nowhere.

What is the significance of all these muddled agendas? The point here is that the simplistic labeling of the nature of the chasm between technologists and their adversaries in generic bumper-sticker terms, and the picturing of two (and only two) starkly divided war camps, hides the many important elements of agreement between us. If we could define our beliefs and our goals in positive, factual terms rather than emotional terms, perhaps we could avoid much needless warfare.

My wish is not to convert anyone to a particular set of beliefs. My goal is more modest. I want to minimize the number of times that my colleagues or I find ourselves talking with persons who share our basic values, and discover that we disagree for totally invalid or irrelevant reasons. There are many subjects on which I disagree with friends for valid and mutually understood reasons. We just have different priorities of the things we value. No problem there. But I wish we could all agree on what the important facts are in a given area, and that requires more than memorizing data. We have to share certain experiences, at least vicariously, and that's what I've tried to offer you in this book.

Suggestions for Technologists

Technologists are being warned from all sides that they may be ruining the planet. There is clearly a basis for this concern, and technologists should not brush off those who express it, just because the accusers may overstate the case or express it in nonscientific or pseudoscientific terms. Technologists have two obligations in responding to these warnings. First to listen—really *listen* and understand what is being said, and be open to the possibility that they may learn something that will help them understand the objector's concern. When the technologist believes the speaker is incapable of understanding the real problem because of a lack of technical expertise, this disdain will show through any polite and flattering rhetoric and block effective dialogue. The other obligation on the technologist is to clarify any technically inaccurate statements of fact that may be critical to defining the problem, and to do so without prejudging the answer. Obviously, carrying out both these responsibilities with tact and competence is a formidable task.

There is also the matter of humility and the broader view. I know from my own upbringing how absolutely true and reasonable it seems that science and rationality are the only way for an intelligent person to know anything for sure, and that these are the final arbiters for judging all matters of importance. Science is unequalled in its ability to analyze a wide range of physical phenomena. But it is not the only way of thinking or of knowing. I know from my more recent experiences that decisions based on hunches, intuition, meditation, prayer and other non-rational modes of thought can often lead to better decisions than pure analysis and reasoning.

Scientists, engineers, and other technologists should accept this perceived contradiction. Surveys show that most of us do operate instinctively, even if we do so surreptitiously. Admiral Rickover's approach to these things was intuitive and instinctive. In discussions over technical issues his arguments were seldom straightforward, rational or even valid. Often, after a bitter argument, events would take an unexpected turn, and Rickover's position would prove to be correct. "Why do you guys fight me on stuff like this?" he would ask. We would try to explain why none of the facts seemed, at the time, to support his conclusion. "But now you tell me I was right. Why am I always right for the wrong reasons?" And here Jack Kyger had the last word: "That's the difference between talent and genius, Captain." Rickover didn't argue with that. But what we were slowly learning is that there is more than one way of thinking and judging.

Most of the contradiction we see is not inherent in the ideas themselves but comes from the labels we assign them—labels that invoke opposite emotional reactions from the debaters. Mark Satin once ran a political newsletter called *New Options,* which discussed new ideas that people were exploring for dealing with various political and sociological problems. He continually got letters from readers, saying such things as "Cancel my subscription. Your so-called new ideas are just warmed-over capitalism in disguise, and I want none of it." Other letters were nearly identical but substituted "socialism" for "capitalism," even when referring to the same article. Once readers found a label they could paste onto an idea, they did not have to think about it. They knew how they felt about the label, and that was enough. (Satin continues this dialogue is his new newsletter, *The Radical Middle.)*

This process of premature labeling seems to me to be at the root of much of our inability to communicate. We aren't willing to turn off our mental critic long enough to hear and think about what the other fellow is saying. And sometimes, instead of taking the form "that's really just *fill-in-the-blank*, and I'm against that," it takes the form of "that's science (or math, or modern art), and I don't understand that stuff." And, again, we turn off our innate ability to understand.

Has Science Killed God?

Many people believe that science has proved that religion is an outmoded superstition, and that all questions concerning the universe and its workings have—or will some day have—materialistic answers. To discuss this matter intelligently, we have to distinguish between two questions. First, what can science tell us that confirms or disproves various specific statements in the Bible (or other sacred writings)? Second, is there scientific evidence for (or against) the existence of any sort of plan, purpose, or intelligence in the formation and operation of the universe?

The first question poses a problem, because many people's belief in God rests on specific, testable factual statements in the Bible or in their own particular sacred text. That's too bad, because scientists have amassed pretty good evidence that the earth does not rest on the back of a large turtle, and that the earth was not created in seven 24-hour days in 4004 B.C. But let's take a more interesting example: Suppose a scientist were to demonstrate that parthogenesis (virgin birth), which has already been demonstrated in frogs and rabbits, was now routinely possible in humans—no miracle needed. Or conversely, that virgin birth was biologically impossible in humans. Would either event deal a serious blow to Christianity (or turn some people away from science)? That's a question that different Christians would answer differently.

Now consider the more important question: Could the universe come into existence and operate without any external direction or purpose? This question would not have been taken seriously by many people until recently; the idea that everything in the universe was created by blind chance would be just too hard to swallow. But within the past few decades, scientists have come up with some incredible examples of how beautiful, complex, and seemingly intelligent behavior can result from the interplay of simple mindless entities. Computer programs following simple fixed rules can create, without human intervention, detailed pictures of various kinds of plants, starting from "seeds," and can simulate evolutionary behavior to a remarkable degree. It is quite amazing how far this can go. Various robotic devices, game-playing machines and decision-making computers can do many "thinking jobs" better than most people.

Many scientists on the cutting edge of biological research are increasingly confident that among the neurons, microtubles and neuro-peptides they will find a purely materialist explanation for Life. In *The Astonishing Hypothesis,* Sir Francis Crick states this view succinctly:

> You, your joys and sorrows, your memories and ambitions, your sense of personal identity and free will, are, in fact, no more than the behavior of a vast assembly of nerve-cells. As Lewis Carroll's Alice might have phrased it: "You're nothing but a pack of neurons."

Biologist William Provine of Cornell University wrote even more bluntly in *The Scientist* (September 5, 1988) an article entitled: "Face It: Science and Religion Are Incompatible!" With confident authority he wrote: "no purposive principles exist in nature … humans die completely with no survival of soul or psyche … no inherent moral or ethical laws exist, nor are there absolute guiding principles for human society … we have no ultimate meaning in life." Another "leading scientist" is quoted as saying "What we know about physics rules out the notion of God existing."

My response was published in the October 17 issue: "Where are the data? What branch of science claims to have solved these questions, which have challenged the greatest thinkers of humankind? In truth, science has no tools to deal with such matters. That is why there are still departments of philosophy and theology in great institutions of learning." All the evidence of science is purely circumstantial; it is not conclusive.

Some scientists are fond of saying that there is absolutely no evidence that God exists. That statement is debatable, of course. But even if true, the absence of evidence is not evidence of absence. Let me give an example: You are in your car, approaching a blind intersection with tall buildings on either side. From where you sit, there is absolutely no evidence of another car approaching the intersection from the side. Are you then safe in assuming that there *is* no car coming? Of course not. That's why there are stop signs.

Atheism as a personal belief is apparently not enough for some scientists; they must proselytize. I agree that scientists should object when religion tries to claim its dogma as science, e.g. when it insists that "creation science" be taught as biology. But I cannot understand or condone the vituperative rhetoric evoked when attempts are made to understand and broaden the relationship between religion and science. When Susan Howatch, a successful novelist, gave $1.5 million to Cambridge University to establish a new graduate seat where advanced-degree students can do research into the complementary nature of science and religion, the scientific establishment exploded. *Nature*, the prestigious international journal of science, castigated Cambridge for stooping so low as to create such an "empty" academic post. Richard Dawkins, a noted zoologist, led a pack of prominent scientists in a scathing attack on the post and on religion: "What has [it] ever said that is of the smallest use to anybody? When has [it] ever said anything that is demonstrably true?... The achievements of theologians don't even mean anything!"

This is not scholarly dissent; it is fear speaking. What are they afraid of?

How Do We Know What We Know?

There is no agreement as to what constitutes knowledge. How do we come to know what we know? The mystic looks within (or to God) to develop core beliefs. If someone says, "God loves me. This I know," you can be sure of two things: that person cannot be reasoned out of that belief, nor can the belief be taught in a classroom. Each person must reach his or her own conclusions in that regard. Scientists, on the other hand, claim that their method of observation and analysis is the only sure way to Truth. I say *claim*, rather than believe, because many scientists admit that some of their most important insights have come in an intuitive flash, without analysis or ratiocination.

So scientists tend to rule out of hand any conclusions based on anything but their own brand of rationality, and they can't understand how anyone could disagree with them. But the rest of the world continues to rely on intuition, or the advice of an older person whom they've grown to trust. Artists have their own way of depicting what's going on in the world, coincidentally coming up with *The China Syndrome* just before the Three Mile Island accident, *The Hot Zone* and *Outbreak* just before the Ebola plague scare, and *Minority Report* just before the Justice Department began locking up people they believed were intending to commit terrorist acts. Lawyers and judges have developed yet another mechanism for arriving at Truth. Philosophers have theirs and theologians theirs. The point is that scientists have to realize that they can speak with authority on only a very limited range of subjects, and outside that range they must try to persuade, not pontificate. Nor can they dictate the terms of the debate.

Science, philosophy, religion, the arts, and the law are all valid ways to search for truth and to describe reality within their own domain, and they need not conflict with each other because they operate on different planes. But we still have to talk and work with each other, from one realm to another.

Making Nuclear Decisions

I am not claiming that nuclear power advocates are always right, or that anti-nuclear activists are always bad guys. I merely ask that you see neither the pro- nor the anti-nuclear spokespersons as selfless public servants, nor yet as greedy pursuers of personal gain at the public expense. There *is* a need for electric power plants, and those who build them and run them rightly think of themselves as working for the public good, particularly when they are working all night through a winter storm to restore light and heat to a stricken neighborhood. It is, of course, essential that these plants not present an undue hazard to the public health or to the environment, and people working toward that end deserve our thanks and support.

We should, however, all watch the conflicts that arise between these groups, evaluate what each side says, judge the validity of the arguments presented, and ask ourselves such questions as: *Are these utility officials really responding to the questions? Are they listening, do they understand, or are they just stonewalling?* Likewise, we should ask of the interveners: *Are they really trying to get a safer plant, or are they just trying to delay it and kill it? Do they know what they're talking about? Should I worry about the problems they raise?* And we should ask of the media: *Why did you quote this character? Does his opinion merit my concern? Have these questions been adequately answered before?* The bottom line is this: *If we don't build this plant, what should we do? Would a coal-fired or an oil-fired plant be any better, all things considered? Is there another alternative, such as solar power, that is really ready to go now? Do we need a new power plant to prevent power outages and brown-outs, or do we already have enough electric power capacity?*

Facts are available. There are reliable people we can talk to. We need not be controlled by pressure groups with hidden agendas or by media selling only fear and sensationalism. It is not beyond the capacity of any intelligent person to reach sound conclusions on these matters. All it takes is a little effort and a lot of patience.

Giving power to The People in the form of the split atom is not sufficient. The real power lies in the people's willingness to inform themselves on the few basic facts and concepts needed to make intelligent choices. A key to accomplishing this is to learn to recognize the signs telling you that you are not applying the same rational criteria to a new idea that makes you uncomfortable that you would apply to another equally outrageous idea that you can live with. You don't have to agree with the new idea, but you should try to understand what it is that makes you uncomfortable with it. There are rational reasons to be concerned about nuclear power just as there are with anything that significantly impacts so many people. But we owe it to ourselves and our neighbors to identify and clarify the basis for those concerns. Thus empowered, we can then apply this knowledge and wisdom to ensure that the physical power is produced and used in a way that maximizes the ability of all people to meet their immediate and long-range needs while protecting and enhancing the fragile planet we share with so many other species great and small. Failure to exercise that power, or giving it away thoughtlessly to politicians or lobbyists, disempowers us all.

Epilogue

Socrates Talks About Nuclear Energy

The public has been told so many things about nuclear energy that are confusing or just plain wrong, that some of the basic physical facts should be clarified. The best way to do this was invented over 2000 years ago by the philosopher Socrates. So I've put this discussion into the form of questions by Socrates, and I'll have his questions answered by a metaphorical Dr. Proh, who sees no serious problems with nuclear technology, and Dr. Kahn, who feels the problems outweigh the advantages. [This Epilogue is based on a piece I wrote for Cosmos 2002, the Cosmos Club journal.]

Socrates: Why is nuclear waste a problem?
Kahn: Because we don't know what to do with it.

S: Why do we have to do something to it?
K: Because it's dangerous.

S: Bicycles and stairs kill people. Does nuclear waste kill or injure people?
K: No, but it can.

S: How can nuclear waste kill?
K: If it leaks into water that may be used for drinking.

S: Is nuclear waste liquid?
K: Sure. There are those huge tanks at Hanford, Washington.

S: Does Hanford store waste from civilian facilities?
Proh: No, virtually none; just weapons wastes.

S: So if we never built any nuclear power plants, it wouldn't change the situation at Hanford.
P: Correct.

S: What form is the waste from power plants?
P: Either spent fuel or miscellaneous waste products. We really shouldn't call spent fuel waste. Only three percent of the fuel has been used; the rest is available for recycle. You wouldn't call a used car waste, if it had been driven, say, 3,000 miles.

S: Are any of these materials liquid?
P: No, the fuel is hard ceramic pellets in metal tubes. The waste is consolidated into a solid—glass, concrete or bitumen. There may be some noble gases, but they are biologically inert and thus no real problem. Even if water were to wash over it, it could not leach out much from glass or metal-clad ceramic.

S: Why then is civilian nuclear waste dangerous?
K: It's radioactive. It gives off dangerous radiation.

S: So we should put it in casks with radiation shielding, right?
P: Well, they do, of course.

S: Can you get dangerous radiation from a nuclear waste cask?
P: No, you'd have to eat the waste.

S: How is that different from a non-radioactive poison?
K: Nuclear waste stays toxic for so long.

S: Doesn't nuclear waste continually *decrease* in toxicity?
K: Yes

S: Stable elements maintain their toxicity undiminished forever. Why is nuclear waste more dangerous?
K: Well, if all U.S. electricity were made from nuclear, the nuclear waste could kill 10 billion people!

S: What does that really mean? Is nuclear waste actually killing 10 billion people?
P: No. It's a hypothetical figure, meaning that the total production is 10 billion times the individual lethal dose. It's like saying that a community swimming pool has enough water to drown a million people. It has no real meaning.

S: How does production of nuclear waste compare with the annual U.S. production of other toxic materials?
P: We produce many common substances with thousands of times greater toxicity. For example, we produce enough chlorine gas each year to kill 400,000 billion people. Then we purify drinking water with it.

S: We don't seem threatened by these, do we?
K: But we keep increasing world's radioactivity, no?
S: Let's see. Where does nuclear waste originate?
P: From the fissionable isotope of uranium.
S: Isn't that naturally radioactive? What is its half-life?
P: Nearly a billion years.
S: So we take a billion-year material and convert it to fission products with mostly shorter half-lives. What's the ultimate effect of that on the earth?
P: In the long run, we make earth less radioactive.

S: Any other problems?
K: Well, there's so much of this waste—thousands of tons of it!

S: How does that volume compare with coal-fired plants, the major competitor to nuclear?
P: A 1000 megawatt coal-fired power plant, supplying all the electricity used by a million people produces 8 million tons of carbon dioxide, which can contribute to global warming; 100 thousand tons of sulfur dioxide, which can cause acid rain and respiratory problems; Nitrogen oxides equivalent to 200,000 automobiles; benzpyrene and other carcinogens; and a quarter million tons of ash containing enough uranium to make several a-bombs. This does not include the mountain tops pushed into valleys to get at Appalachian coal seams

By contrast, a nuclear reactor generating the same amount of electricity produces two cubic meters of waste, which can all be sealed in containers and controlled, not dispersed into the environment. All waste from 40 years operation is stored at the plant. They could store another 40 years worth.

S: If all your electricity were produced by nuclear power, how much waste would that represent?
P: You could store your life's share in a corner of your basement.

S: Why don't we put the nuclear waste into the sea? Would it despoil the whole ocean?
P: No. The ocean's natural radioactivity would completely overshadow it. You could detect it only with special instrumentation that discriminates one isotope from another. Rivers continuously dump more radioactivity into the ocean than we create with all our nuclear power plants. We could put it in drums and drop them into the deep ocean clay, where they'd be isolated for millennia.

S: What if we did nothing about this problem for several years?
P: Even anti-nuclear activists agree there's no safety problem. The few sites shutting down would need the fuel casks sent elsewhere. Some other sites might need to increase their storage capacity, but that's not difficult. Some states have laws limiting fuel storage, but that is a man-made problem; it could be fixed.

S: Is that a multi-billion dollar task?
P: No. There are many government or private sites nearby that could store the casks for a few decades.

K: But then what would we do?
P: Nuclear waste contains many valuable products. We will ultimately want to recover those, as well as the unspent fuel. If this source of energy were in the form of oil, we'd be ready to sacrifice a generation to protect it. As coal, we'd destroy pristine mountains to get it. But here we have it, already refined, within our borders, ready to use. Right now it's cheaper to use uranium ore, but the spent fuel is there for the future.

K: But radiation is carcinogenic, right? It's always a danger.
P: No. In fact, radiation in small doses is beneficial. It's like selenium or other trace elements in your vitamin pills. In large doses they're deadly poisons. In small doses they're actually nutrients.

S: How can that be? Isn't nutrient the exact opposite of poison? Can the same substance be both?
P: Paracelsus said in 1540 that nothing is poison, but the dose makes it so. That's how vaccination works. Radiation acts the same way. Large amounts are poisonous, small amounts are beneficial. In fact, there are experiments demonstrating that reducing the natural radiation background causes organisms to get sick and die. And people who live where natural radiation is high generally live longer and have less cancer.

K: Why is that? Doesn't each cell damaged by radiation create a potential cancer? Twice as much radiation causes twice as many damaged cells. That's got to double your risk of cancer, no?

S: Doesn't each flu germ entering your body create a potential disease?
K: Yes, but...

S: Then do you conclude that the best way to fight disease is to keep washing your hands, wipe off doorknobs, avoid shaking hands—to minimize contact with germs, as some germophobes do?
P: That's not what doctors recommend.

S: What do doctors recommend? And why?
P: They say keep warm, eat nutritious food, exercise, to keep your body healthy. So the number of germs in our bodies is less relevant than the state of our immune systems and other defenses. If you keep healthy, your body will take care of the germs.

S: Does radiation work that way?
K: I don't think so. Radiation damages cells, and that's how cancers start.

S: Are cells in your body ever damaged by events other than radiation exposure?
P: Oh, yes. Normal metabolism damages hundreds of millions of cells for each one damaged by background radiation.

S: Is it the same kind of damage?
P: Not exactly. Radiation damage is harder to repair. But we know how much harder, and the net result is that even after repair, metabolism still leaves several million more damaged cells than radiation does. The number of damaged cells is not the critical factor, despite how some people argue. High dose radiation kills, not by damaging more cells, but by degrading the defense system.

S: Then how does low-dose radiation affect the body beneficially?
P: When it stimulates the defense system, it enhances repair and replacement of not only the few radiation-damaged cells, but also the very much larger number of metabolically-damaged cells.

S: One fact that complicates discussions of radioactivity is the presence of natural radioactivity. We've had congressmen urging that we "get it down to zero." How do the natural radioactivities compare with some regulatory limits?

P: To answer this, I have to explain what the numbers mean. Radioactivity measures how intense a radiation source is (just as luminosity measures how bright a light source is). The amount of radiation one gets depends on the strength of the source, its distance from the receiver, and whether any shielding is present.

We measure radioactivity in curies (named after Marie Curie, the discoverer of radium). One curie is the amount of radioactivity possessed by one gram of pure radium (1/28 of an ounce). We usually encounter much less than one curie, so we measure lesser amounts in picocuries (millionths of a millionth of a curie). In one picocurie, only about 2 atoms per minute are decaying and giving off radiation. The radioactivity of a liquid is measured in picocuries per liter (a liter is a little more than a quart). (Some years ago, equivalent metric units were defined, but U.S. regulations are still set in curies.)

So let me give you some examples in picocuries per liter. The proposed EPA radium limit on tap water is 5. The natural level of river water is from 10 to 100. Natural seawater is 300. Whiskey is 1,200, milk is 1,400 and salad oil is 5,000. And natural radon in much of the world's drinking water is 30,000, and some health spa waters are as high as 300,000. There is considerable evidence that these natural levels of radioactivity are not harmful and are probably beneficial.

S: So it appears we are protecting people against a truly non-existent hazard! How about terrorism? We read some really frightening possibilities. Plutonium, we're told, is the deadliest poison known? Is this true?
P: No. You can hold it in your bare hands. Spoonful for spoonful, it is about as toxic as caffeine. When physicist Bernard Cohen was told that he and other scientists were not interesting interview subjects, he offered to eat on-camera as much plutonium as Ralph Nader would eat caffeine. But the interviewer said that would be cheap exhibitionism.

S: So where does plutonium get this reputation?
K: It is considerably more lethal if inhaled.

S: Then the scenario of putting plutonium into a public ventilator would create a real disaster?
P: No, plutonium is very heavy and is extremely hard to keep suspended in air. It wouldn't work well. During all the decades we have been handling plutonium in tonnage lots there has never been a death from plutonium toxicity. Even after dispersing some 5 to 7 tons of it into the air during 1,000 weapons tests.

S: What about terrorist airplanes? Can an airliner fly through several feet of steel-lined reinforced concrete?

P: No. The plane would either slide off the curved surface or crush like an eggshell outside. Any jet fuel would burn harmlessly outside. The size of the plane is relatively unimportant, since the plane structure collapses on itself, absorbing most of the impact energy, and only the engines pose a penetration potential.

P: In 1988, a full-sized (unmanned) Phantom F-4 fighter plane was driven by rockets at 480 miles per hour into a simulated containment wall section. These instrumented tests confirmed analysis: the body of the plane crushed against the outside, penetrating less than an inch. The engine shaft penetrated less than two inches. It is clear that an airplane cannot fly through such a wall. It is true but largely irrelevant that the idea that a suicide pilot might undertake such an attack was not considered prior to September 11.

K: But if terrorists got inside the plant with explosives, couldn't they, as speculated in the papers, create a disaster like that burning reactor accident in Chernobyl in 1986, with tens of thousands of deaths?

S: What is the worst that could happen?

P: No credible sequence of events involving a U.S. reactor could lead to tens of thousands of deaths. Even Chernobyl caused no deaths to the public, even without containment and without evacuation for the critical first days. The U.N. scientific report (UNSCEAR 2000) reports no other deaths than the 30 workers and firefighters in the plant. The 2,000 thyroid cancers were said to be 97% curable and probably due to intensive screening, since they do not correlate with radiation dose. An American reactor meltdown, at worst, would be more like Three Mile Island where there were no significant health or environmental effects whatsoever, even to plant workers.

K: But wasn't that due primarily to the intact containment structure that held in the fission products? What if the containment was breached?

P: Studies after the accident showed that nearly all of the harmful fission products dissolved in the water and condensed out on the inside containment surfaces. Even if containment had been severely breached, little radioactivity would have escaped. Few, if any, persons would have been harmed. Tons of molten reactor sitting on the 5-inch-thick reactor vessel bottom did not even penetrate the 5/16-inch cladding. So much for the dreaded *China Syndrome*!

K: What if terrorists got hold of a spent-fuel shipping cask?
S: What is the worst that could lead to?

P: There is nothing one can do to a spent fuel shipping cask that could lead to a significant public hazard. Despite frightening claims about the hazards of what fear-mongers call "Mobile Chernobyls," spent fuel casks pose no significant public hazard. They cannot "go critical" like a reactor or detonate like a bomb. None of the radioactivity is in liquid form. It is solid ceramic pellets, metal clad. For more than 30 years over 5,000 fuel assemblies have already been shipped. Despite a few serious traffic accidents, not a single radiation release has occurred. The fuel in these casks is always cooled for several years prior to shipment, so the short-lived activity and the decay heat production have died down. The shipping casks themselves are virtually indestructible. To be certified for shipping, a cask must be able to withstand a 30-foot drop onto its edge, a 40-inch drop onto a puncture bar, a 1475°F fire for 30 minutes, immersion under 50 feet of water for 8 hours. Further crash tests have involved a tractor trailer carrying the cask hitting a concrete wall at 84mph, a locomotive hitting the cask broadside at 80mph, crash at 80mph followed by 125 minutes completely engulfing jet fuel fire, and a drop test from a helicopter so that the cask buried itself more than 4 feet in the hard-packed ground. In addition, casks have been tested with high-tech anti-tank explosive charges. Only in this last case was the cask breached, but even then the result of scattering a few chunks of spent fuel on the ground could not create a serious public hazard. There is no mechanism to disperse the radioactivity in an ingestible or respirable form, over a significant distance. At the worst, only a very few people would get some radiation doses and these would not be life-threatening.

K: But there's still the "dirty bomb." A terrorist wraps radioactive material around an ordinary explosive and supposedly spreads death and destruction.

S: Is this a real threat?
P: No. It is completely ineffective. Many tons of shielding would be required to permit handling by deliverers. The radioactive ceramic scatters only a short distance. The noble gases are biologically inert. Little air, water or land contamination results. There would be few if any casualties beyond the reach of the explosion. This is not a credible weapon.

S: So are you saying we need not be careful in dealing with nuclear technologies?
P: No, of course not. We have taken extraordinary precautions, and consequently no one has been killed or even seriously injured by American-type nuclear power plants or their waste products. But this has had the perverse effect of scaring people into thinking we must have an unimaginably dangerous beast, to justify such extreme precautions. That's why it is good we talked about how the laws of nature and the physical properties of materials prevent any major public hazard, in any credible circumstance we can think of.

K: Wait a minute! I've read many times about some guy, usually a kid, getting burned by some radiation source. What about that?
P: Radiation now has thousands of industrial and commercial uses, some of which used to be done by x ray tubes. Just as people are occasionally hurt by inexcusably careless use of x ray tubes, so one can steal a radiographic source, take it out of its shield and carry it in his pocket, or play unknowingly with radioactive power and make a mess. But these injuries result from illicit use of industrial equipment and generally affect only the miscreants and sometimes their families or associates. While unfortunate, they are in the same category as accidents involving stolen tractors, police cars, or medicinal narcotics. We don't condemn the legitimate use of such things; we just tighten up on security. Such incidents are no more frequent or more damaging with radiation devices than with many other types of equipment.

S: Well, I think we've covered about as much information as one can comfortably absorb in one session. Let's get together again soon, shall we?

INDEX OF SUBJECTS

A

B

C

D

E

F

G

H

I

J

K

L

M

N

O

P

R

S

T

U

V

W

Z

"Presents in vivid, human terms many of the young scientists and engineers who first harnessed this primal force, and the extraordinary times and environment in which they worked and lived...an enlightening and fascinating account."

From the Foreword by the late ***Dr. GLENN T. SEABORG****, Nobel Laureate, Co-discoverer of plutonium, Chairman, U.S. Atomic Energy Commission, 1961-71*

"A unique contribution...I don't know of any other book that covers the same ground—which was ground zero for the evolution of this important and controversial technology ... It doesn't hurt that you're an engaging storyteller and that you were present at the creation."

RICHARD RHODES*, Independent journalist and historian, winner of a Pulitzer Prize, a National Book Award, a National Book Critics Circle Award, and a History of Science Society Award*

"His prose disproves the prevalent belief that no engineer can write. The sentences flow gracefully, with a feel for the rhythm of English that many who make a profession of writing would envy."

CONNIE BUCHANAN*, Editor of Tom Clancy's "Hunt for Red October"*

"Ted Rockwell takes countless disaggregated fragments of history and welds them together in a fascinating story that all Americans can read, understand, and enjoy. A gifted scientist, engineer, visionary, and author, Ted has been a front-line player throughout this Age—eminently qualified to tell us the true story and set the images straight."

Admiral James Watkins*, USN (ret), Chief of Naval Operations, 1982-86; Secretary of Energy, 1989-93.*

Printed in the United States
40492LVS00004B/38